Artificial Intelligence Annual Volume 2022

Edited by Marco Antonio Aceves Fernandez and Carlos M. Travieso-Gonzalez

Published in London, United Kingdom

Artificial Intelligence Annual Volume 2022
http://dx.doi.org/10.5772/intechopen.109246
Edited by Marco Antonio Aceves Fernandez and Carlos M. Travieso-Gonzalez

Contributors
Jing-Sin Liu, Hc Liao, Han-Jung Chou, Iqramul Haq, Md. Ismail Hossain, Md. Moshiur Rahman, Md. Injamul Haq Methun, Ashis Talukder, Md. Jakaria Habib, Md. Sanwar Hossain, Ewa J. J. Kleczyk, Michele Bennett, Karin Hayes, Rajesh Mehta, Yalın Baştanlar, Semih Orhan, Nabil Ajali, Carlos M. Travieso-Gonzalez, Kishore Kumar Kamarajugadda, Pavani Movva, Suxia Cui, Soamar Homsi, Veeramani Sonai, Indira Bharathi

First published in London, United Kingdom, 2022 by IntechOpen
IntechOpen is the global imprint of INTECHOPEN LIMITED, registered in England and Wales, registration number: 11086078, 5 Princes Gate Court, London, SW7 2QJ, United Kingdom

British Library Cataloguing-in-Publication Data
A catalogue record for this book is available from the British Library

Additional hard and PDF copies can be obtained from orders@intechopen.com

Artificial Intelligence Annual Volume 2022
Edited by Marco Antonio Aceves Fernandez and Carlos M. Travieso-Gonzalez
p. cm.

This title is part of the Artificial Intelligence Book Series, Volume 12
Series Editor: Andries Engelbrecht

Print ISBN 978-1-83768-946-0
Online ISBN 978-1-83768-947-7
eBook (PDF) ISBN 978-1-83768-948-4
ISSN 2633-1403

IntechOpen Book Series

Artificial Intelligence

Volume 12

Aims and Scope of the Series

Artificial Intelligence (AI) is a rapidly developing multidisciplinary research area that aims to solve increasingly complex problems. In today's highly integrated world, AI promises to become a robust and powerful means for obtaining solutions to previously unsolvable problems. This Series is intended for researchers and students alike interested in this fascinating field and its many applications.

Meet the Series Editor

Andries Engelbrecht received the Masters and Ph.D. degrees in Computer Science from the University of Stellenbosch, South Africa, in 1994 and 1999 respectively. He is currently appointed as the Voigt Chair in Data Science in the Department of Industrial Engineering, with a joint appointment as Professor in the Computer Science Division, Stellenbosch University. Prior to his appointment at Stellenbosch University, he has been at the University of Pretoria, Department of Computer Science (1998-2018), where he was appointed as South Africa Research Chair in Artifical Intelligence (2007-2018), the head of the Department of Computer Science (2008-2017), and Director of the Institute for Big Data and Data Science (2017-2018). In addition to a number of research articles, he has written two books, Computational Intelligence: An Introduction and Fundamentals of Computational Swarm Intelligence.

Meet the Topic Editors

Dr. Marco Antonio Aceves Fernandez obtained his B.Sc. (Eng.) in Telematics from the Universidad de Colima, Mexico. He obtained both his M.Sc. and Ph.D. from the University of Liverpool, England, in the field of Intelligent Systems. He is a full professor at the Universidad Autonoma de Queretaro, Mexico, and a member of the National System of Researchers (SNI) since 2009. Dr. Aceves Fernandez has published more than 80 research papers as well as a number of book chapters and congress papers. He has contributed to more than 20 funded research projects, both academic and industrial, in the area of artificial intelligence, ranging from environmental, biomedical, automotive, aviation, consumer, and robotics applications. He is also an Honorary President of the National Association of Embedded Systems (AMESE), a senior member of the IEEE, and a board member of many institutions. His research interests include intelligent and embedded systems.

Carlos M. Travieso-Gonzalez received his M.Sc. degree in Telecommunication Engineering at the Polytechnic University of Catalonia (UPC), Spain in 1997, and his Ph.D. degree in 2002 at the University of Las Palmas de Gran Canaria (ULPGC-Spain). He is a full professor of signal processing and pattern recognition and a Head of the Signals and Communications Department at ULPGC, teaching from 2001 on subjects on signal processing and learning theory. His research lines are biometrics, biomedical signals and images, data mining, classification system, signal and image processing, machine learning, and environmental intelligence. He has participated in 52 international and Spanish research projects, some of them as a lead researcher. He is co-author of four books, co-editor of 27 proceedings books, guest editor for eight JCR-ISI international journals, and author of up to 24 book chapters. He has over 450 papers published in international journals and conferences (81 of them indexed in JCR – ISI - Web of Science) and has published seven patents in the Spanish Patent and Trademark Office. Dr. Travieso-Gonzalez has been a supervisor on eight Ph.D. theses (11 more under supervision), and 130 master theses. He is the founder of the IEEE IWOBI Conference Series and the President of its Steering Committee, as well as the founder of both the InnoEducaTIC and APPIS Conference Series. He is an evaluator of the project proposals for the European Union (H2020), Medical Research Council (MRC, UK), Spanish Government (ANECA, Spain), Research National Agency (ANR, France), DAAD (Germany), Argentinian Government, and the Colombian Institutions. He won the "Catedra Telefonica" Awards in Modality of Knowledge Transfer, 2017, 2018, and 2019 editions, and Award in Modality of COVID Research in 2020.

Contents

Preface

Since the times of the earliest humans to populate the earth, we have gradually tried to understand and control the world around us. In an attempt to understand such phenomena, humans started to make predictions to various degrees. For instance, humans started to make predictions about the planets' motions, eclipses, rainfall cycles, or the periodicity of certain diseases. However, in the last few decades, the complexity of carrying out predictions has exceeded our abilities to predict.

Fortunately, the dawn of electronic computers is profoundly increasing our abilities to predict nature. However, the problems we are facing now are far more complex than the problems we faced a century ago.

The ability of these machines to demonstrate advanced cognitive skills in making decisions, learning, perceiving the environment, predicting certain behaviors, or processing written or spoken languages, among other skills, makes this discipline of paramount importance in today's world.

I hope that this work is of interest to students and researchers alike, as I did my best to comprise quality research contributions with several different applications.

Marco Antonio Aceves Fernandez
Universidad Autonoma de Queretaro,
Queretaro, Mexico
Topic Editor: Machine Learning and Data Mining

Artificial Intelligence has been a very important topic over the last period due to its use in companies, business, and real life because more and more data is generated each time. The automatic interpretation of big data is based on the extraction of patterns, and the field of Artificial Intelligence has a great role in extracting information and making decisions.

In particular, this book presents some of the contemporary and relevant topics in Artificial Intelligence, providing machine learning approaches, deep learning approaches, knowledge-based recognition, case studies and emerging technologies, and applications using Artificial Intelligence. Innovative advances from this field have been included in order to show the reader the newly researched and developed approaches.

Carlos M. Travieso-Gonzalez
University of Las Palmas de Gran Canaria,
Gran Canaria, Spain
Topic Editor: Applied Intelligence

Section 1

Machine Learning and Data Mining

Chapter 1

Machine Learning Algorithm-Based Contraceptive Practice among Ever-Married Women in Bangladesh: A Hierarchical Machine Learning Classification Approach

Iqramul Haq, Md. Ismail Hossain, Md. Moshiur Rahman, Md. Injamul Haq Methun, Ashis Talukder, Md. Jakaria Habib and Md. Sanwar Hossain

Abstract

Contraception enables women to exercise their human right to choose the number and spacing of their children. The present study identified the best model selection procedure and predicted contraceptive practice among women aged 15–49 years in the context of Bangladesh. The required information was collected through a well-known nationally representative secondary dataset, the Bangladesh Demographic and Health Survey (BDHS), 2014. To identify the best model, we applied a hierarchical logistic regression classifier in the machine learning process. Seven well-known ML algorithms, such as logistic regression (LR), random forest (RF), naïve Bayes (NB), least absolute shrinkage and selection operation (LASSO), classification trees (CT), AdaBoost, and neural network (NN) were applied to predict contraceptive practice. The validity computation findings showed that the highest accuracy of 79.34% was achieved by the NN method. According to the values obtained from the ROC, NN (AUC = 86.90%) is considered the best method for this study. Moreover, NN (Cohen's kappa statistic = 0.5626) shows the most extreme discriminative ability. From our research, we suggest using the artificial neural network technique to predict contraceptive use among Bangladeshi women. Our results can help researchers when trying to predict contraceptive practice.

Keywords: contraceptive, machine learning algorithms, LASSO, NN, hierarchical

1. Introduction

Family planning is indispensable in facilitating the prosperity and autonomy of women, their families, and their communities. Contraceptive choices, maternal and newborn health care, sexually transmitted infections, and sexual health are the main concepts of reproductive health [1]. The states agreed in 2001 that among the Millennium Development Goals (MDGs), target 5b was called for by 2015 for universal access to reproductive health. Global contraceptive prevalence is 64% (41% in low-income countries) and the global unmet need for family planning is 12% (22% in low-income countries) as reported at the end of the MDGs period. Sustainable Development Goals (SDGs) targets 3.7 and 5.6 call for universal access to sexual and reproductive health care services and sexual and sexual and reproductive health and reproductive rights, respectively [2, 3].

It has been calculated that maternal mortality has been reduced globally by 30% by the increase in contraceptive use [4]. Unintended pregnancies, pregnancy spacing, and reducing high-risk pregnancies are the consequences of contraceptive use [5–7]. Current studies show that every year, contraceptive use could reduce nearly 230 million births by stopping unwanted pregnancies [8]. As a result, the use of contraception improves the health of women and their children [6, 9]. However, the prevalence of contraceptive practice varied between 11.3% and 72.1% in different countries, namely Mozambique, 11.3%, Ghana, 21.5%, Bangladesh (modern method), 54.0%, and Sweden, 72.1% [9–12].

Previous research has shown that various variables are significantly associated with contraceptive use, such as maternal age, maternal and husband's educational level, wealth status, maternal age at first marriage, and so on [11, 13]. Through the promotion of family planning, appropriate diagnostics, and interventions, the prevalence of contraceptive use is increasing. Popular statistical methods (binary logistic regression) have been applied to determine important indicators of contraceptive use among women. But the main goal is to predict contraceptive practice among women aged between 15 and 49 in Bangladesh. Machine learning is a scientific method that can build models for prediction purposes. According to the research, traditional statistical procedures were shown to be ineffective in this form of modeling. Machine learning approaches have long been shown to be more successful and promising in handling a variety of complicated and nonlinear issues [14–16].

However, not many studies have explored machine learning methods to develop predictive models for studying contraceptive methods. Therefore, various well-known machine learning algorithms were applied to predict contraceptive practices among 15–49-year-old women in Bangladesh in this study. Before prediction, we applied a Hierarchical Logistic Regression classifier in machine learning approaches that were used to select potential risk factors associated with the contraceptive practice of women. To our best knowledge, the originality of the study is that it is almost new in the field of machine learning classifier approach in the contraceptive practice of Bangladesh context, for the first time using such methods, which will assist future data scientists.

2. Methods

2.1 Data source

In this study, the necessary information has been extracted from a representative secondary national data set, the Bangladesh Demographic and Health Survey (BDHS),

2014. This survey was carried out through a joint effort of the National Institute of Population Research and Training (Bangladesh), Mitra Associates (Bangladesh), and ICF International (USA).

The entire list of enumeration areas (EAs) that encompasses the entire country, provided by the Bangladesh Bureau of Statistics (BBS) for the 2011 population and housing census of the People's Republic of Bangladesh, served as the sampling frame for the 2014 BDHS. An EA was a geographical zone with an average of 120 households. The survey uses a two-stage stratified sampling process that includes information on the EA region, residence (urban or rural), and the number of residential households counted. Viable interviews were conducted in 98% of the selected households (out of 17,989 total). For this study, 17,863 ever-married women aged 15–49 years were included in the final analysis. Note that to learn more about the detailed sampling procedure of the 2014 BDHS, see the final published report of the survey [17].

2.2 Dependent variable

Since the main purpose of this study was to predict contraception practice among women aged 15–49 years, the response variable was "current contraception use", which was classified as "Yes or No". If the respondent currently utilizes a contraceptive method, she falls into the "Yes" group, otherwise, she falls into the "No" group.

2.3 Independent variables

Besides the response variable, a set of 21 demographic and socioeconomic risk factors were included in the analysis, which was associated with contraceptive practice and considered predictor variables. Several studies found that demographic and socioeconomic characteristics such as current age, division, religion, residence, respondent's working status, FP media exposure, age at first marriage, currently breastfeeding, wealth status, women's education, husband's education, child ever born, number of living children, ideal number of children, fertility preference, marital status, and decision making for using contraception are potential risk factors that determine contraception practice among women [10, 11, 18–24]. The list of independent variables and their measures are presented in **Table 1**.

No.	Variables	Measures
1	Women current age (years)	15–19, 20–24, 25–29, 30–34, 35–39, 40–44, 45–49
2	Division	Barisal, Chittagong, Dhaka, Khulna, Rajshahi, Rangpur, Sylhet
3	Religion	Islam, other
4	Sex of household head	Male, female
5	Residence	Urban, rural
6	Respondent working status	No, yes
7	Family planning (FP) media exposure	No, yes
8	Age at first marriage	<18, 18+

No.	Variables	Measures
9	Currently breastfeeding	No, yes
10	Currently amenorrhoeic	No, yes
11	Currently abstaining	No, yes
12	Wealth status	Poor, middle, rich
13	Women education	No education, primary education, secondary+
14	Husband education	No education, primary education, secondary+
15	Sexually transmitted infection (STI)	No, yes
16	Children ever born	0–1, 2–3, 4+
17	Number of living children	None, 1–2, 3+
18	Ideal number of children	0–1, 2–3, 4+
19	Fertility preference	No more, have another, undecided, declared infecund, sterilized
20	Marital Status	Married, others
21	Decision making for using contraception	Respondent, others

Table 1.
Description of independent variables.

2.4 Statistical analysis

The frequency distribution was used to describe the background characteristics of the respondents. In this study, we developed a Hierarchical Logistic Regression classifier in machine learning approaches that were used to select potential risk factors related to the contraceptive practice of women in Bangladesh by using the largest value of AUC ($p < 0.05$). One of the procedures for enhancing the performance of machine learning is hierarchical learning, which is inspired by human learning [25]. The DeLong test is an extensively used test to compare the difference between two AUCs [26]. That model was significant, with the largest AUC value, and was considered the final model in this analysis. The steps are depicted in **Figure 1**.

To meet the objective of the study, we fitted numerous numbers of model where the full model is denoted by $\boldsymbol{M_i}$ (where $i = 21$) using Hierarchical Logistic Regression classifier in the Machine Learning Process. The steps are described below:

Step 1: Consider jth model defined as $\boldsymbol{M_j}$ $(j = 1, 2, 3, \dots, i)$ which is consist of j predictors. Thus, the initial model was named Model1 and defined as $\boldsymbol{M_1}$ where $(j = 1)$, then fit the model $\boldsymbol{M_1}$ by using machine learning logistic classifier (MLLC).

Step 2: Adding a variable in the previous model and defined as $\boldsymbol{M_{j+1}}$ and again also fit model $\boldsymbol{M_{j+1}}$ by using MLLC approach.

Step 3: Identify the best model by using Delong's Test, which is considered the largest area under the curve at a 5% level of significance.

Step 4: If $\boldsymbol{M_{j+1}} > \boldsymbol{M_j}$ based on AUC at 5% level of significance, then Model $\boldsymbol{M_{j+1}}$ has a significantly different AUC from Model $\boldsymbol{M_j}$ with $p < 0.05$. In this case, the best model was considered as $\boldsymbol{M_{j+1}}$, otherwise the model was $\boldsymbol{M_j}$.

Step 5: The process is repeated successively until the desired number of risk factors/features are identified.

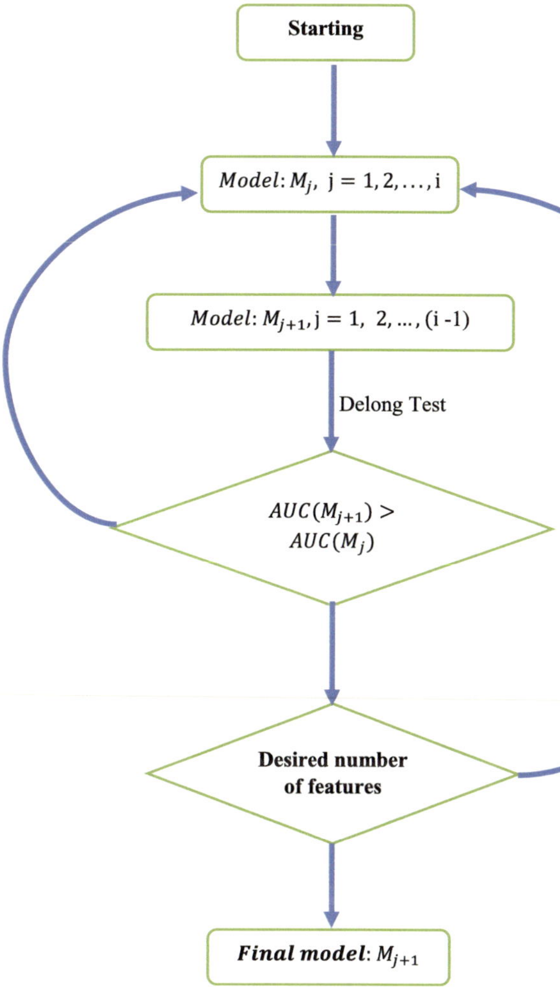

Figure 1.
Flow diagram for hierarchical logistic regression classifier in the machine learning process.

After selecting the final model, we applied the 7 most popular machine learning classifiers to predict contraceptive practice among ever-married women aged 15–49 in Bangladesh. In this study, we used seven different popular ML algorithms (Logistic Regression (LR), Random Forest (RF), Naïve Bayes (NB), Least Absolute Shrinkage and Selection Operation (LASSO), Classification Trees (CT), AdaBoost, and Neural Network (NN)). A detailed description of the algorithms used is available in the literature [27–32].

The Statistical Package for Social Science (SPSS) version 25 and R version 4.0.0 software were used for data management and analysis.

2.5 Proposed approach

Data from ever-married women aged 15–49 was used in this study. Only ever-married women aged 15–49 was considered for the final analysis based on this criterion. Then, apply data preparation methods; for example, first find out missing data from the overall dataset. It is well known that the main drawbacks of missing information in a dataset are the reduced statistical power (because it reduces the number of samples n, the estimates will have larger standard errors). The main disadvantages of missing data in a dataset are statistical power reductions, which are well-known (because it reduces the number of samples n, the estimates will have larger standard errors). There are numerous imputation methods for imputing missing values nowadays, including direct deletion, mode imputation, hot-deck imputation, and so on [33]. A lower threshold of 5% missingness has been suggested in the literature [34]. We utilized the direct deletion method because this study had a low rate of missing values, which means we removed all missing values from the data set and conducted the analysis using the entire data set. The next step after missing value processing is to normalize/standardize the variables, which is useful when the data distribution is unknown. As a result, normalization is not required for any machine learning approach, especially in categorical data. Finally, all machine learning classifiers included in this study were performed on 70% of the respondents in each group (training data set, n = 12,504) and acquired by the remaining 30% (test data set, n = 5358). All models were trained to support 10-fold cross-validation. On the training set, we performed 10-fold cross-validation, and on the testing set, we estimated performance. The results of the development of the seven machine learning classifiers are depicted in **Figure 2**.

2.6 Model evaluation

We used the following criteria to evaluate the ML algorithms' performance: confusion matrix, receiver operating characteristic (ROC), and the area under that curve (AUC). Generally, a confusion matrix has four possible prediction outcomes, such as TR = true positives, TN = true negatives, FP = false positives, and FN = false negatives. Several performance measures, including accuracy, precision, recall, and the F1 score, are usually calculated using these four potential outcomes to assess the classifier. The ROC curves have been calculated by utilizing the predicted outcomes as well as the true outcomes. To examine the ML algorithms' discriminating powers, the AUC of the ROC has been averaged for the test data sets [35]. Theoretically, the AUC should be between 0 and 1, with 1 being the most extreme value for an ideal classifier. Since the usual lower bound for random classification is 0.5, an AUC greater than 0.5 has at least some capacity to separate between cases and non-cases [36]. In addition to these measures, we also used Cohen's kappa statistic, which is a better measure to examine the agreement between two raters. It is calculated by utilizing the predicted and the actual classifications in a data set. The value of Cohen's kappa statistic is 1.

3. Results

3.1 Sociodemographic characteristics of women

Table 2 shows the percentage distribution of women according to the selected socio-demographic characteristics of Bangladesh. The majority of women (19%) are between the ages of 25 and 29. The majority of them (35%) are from the Dhaka division, Muslims (90%), living in male-headed households (89%), and in the rural areas (72%). In terms of working status, slightly more than two-thirds (67%) of women are not currently involved in any kind of income-generating activities, and 80% of them do not have any media exposure. The majority of women (77% of them)

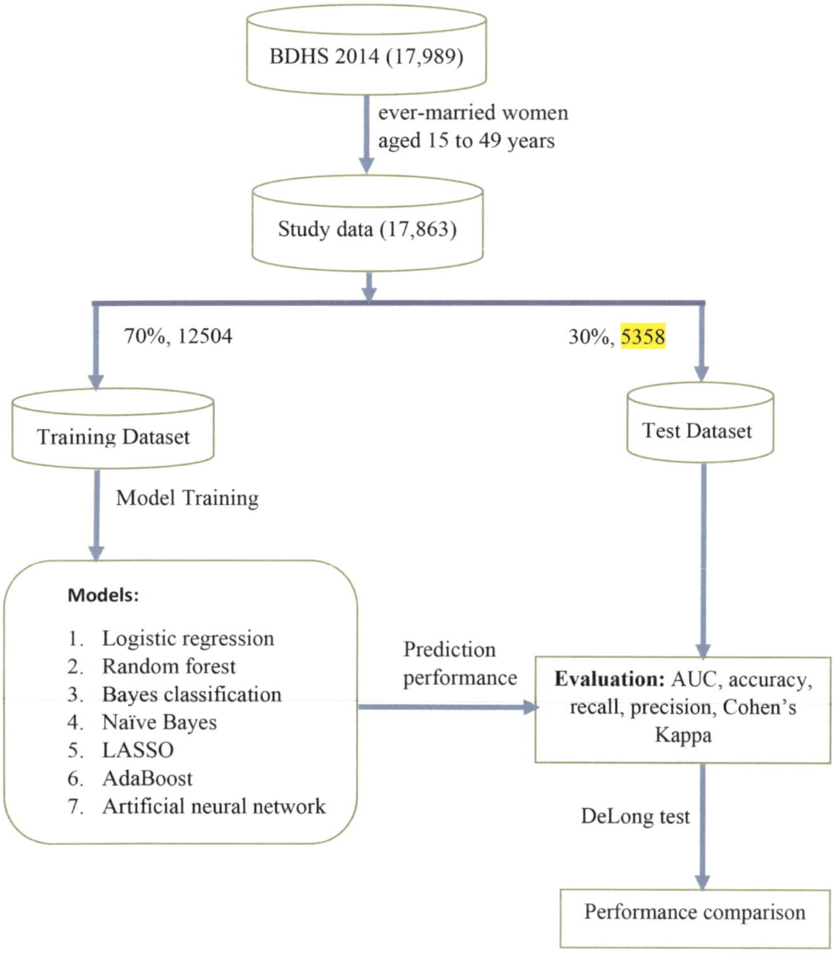

Figure 2.
Flow chart of the development of the seven machine learning classifiers.

Characteristics	Sample women	
	No.	**%**
Women current age (years)		
15–19	2029	11.4
20–24	3224	18.0
25–29	3390	19.0
30–34	3047	17.1
35–39	2315	13.0
40–44	2092	11.7
45–49	1766	9.9
Division		
Barisal	1111	6.2

Characteristics	Sample women	
	No.	%
Chittagong	3301	18.5
Dhaka	6223	34.8
Khulna	1838	10.3
Rajshahi	2103	11.8
Rangpur	2056	11.5
Sylhet	1232	6.9
Religion		
Islam	16,096	90.1
Other	1767	9.9
Sex of household head		
Male	15,854	88.8
Female	2009	11.2
Residence		
Urban	5047	28.3
Rural	12,816	71.7
Respondent working status		
No	11,947	66.9
Yes	5912	33.1
Family planning (FP) media exposure		
No	14,316	80.1
Yes	3547	19.9
Age at first marriage		
<18	13,657	76.5
18+	4206	23.5
Currently breastfeeding		
No	14,033	78.6
Yes	3830	21.4
Currently amenorrhoeic		
No	17,054	95.5
Yes	809	4.5
Currently abstaining		
No	17,341	97.1
Yes	522	2.9
Wealth status		
Poor	6767	37.9
Middle	3560	19.9

Characteristics	**Sample women**	
	No.	**%**
Rich	7536	42.2
Women education		
No education	4455	24.9
Primary	5209	29.2
Secondary +	8199	45.9
Husband education		
No education	5189	29.0
Primary	4289	27.3
Secondary+	7795	43.6
Sexually transmitted infection (STI)		
No	11,947	66.9
Yes	5912	33.1
Children ever born		
0–1	5670	31.7
2–3	8139	45.6
4+	4054	22.7
Number of living children		
None	1814	10.2
1–2	9478	53.1
3+	6571	36.8
Ideal number of children		
0–1	1127	6.3
2–3	15,308	85.7
4+	1429	8.0
Fertility preference		
No more	9555	56.7
Have another	5293	31.4
Undecided	462	2.7
Declared infecund	561	3.3
Sterilized	986	5.8
Marital Status		
Married	16,858	94.4
Others	1005	5.6
Decision making for using contraception		
Respondent	1515	8.5
Others	16,348	91.5.5

Characteristics	Sample women	
	No.	%
Contraception use status		
Using	10,527	58.9
Not Using	7336	41.1

Table 2.
Percentage distribution of ever- married women age between 15 and 49 by selected socio-demographic characteristics.

married before their 18th birthday, and 79% of them were not breastfeeding their children at the time of the survey. The findings also show that around 96 to 97% of women are not amenorrheic (96%) or abstaining (97%). In terms of wealth status, 42% of the women were from rich families. Approximately half of the women (46%) had secondary or higher education. The majority of the husbands (44%) had a secondary or higher level of education. The number of women who knew about sexually transmitted infections (STIs) was found to be 67%. The majority of women (46%) have had 2–3 children, while 53% have 1–2 living children. The ideal number of children was 2–3 (86%) and more than half (57%) of the women were not interested in having another child. The vast majority of women are currently married (94%), and only 9% can make the decision to use a contraception method on their own. Regarding contraception use, according to the 2014 BDHS, 58.9% of women used it.

3.2 Create model

In the initial step of the analysis, we applied hierarchal logistic regression to select the final model. Here, each variable was considered as one model. We added a potential risk factor (variable) to the previous model that was considered a new model in this analysis (**Table 3**). For example, in the initial model M_1 we considered (arbitrary) respondent age, M_1+*Division* was considered as M_2. Similarly, we consider another model by adding a variable to the previous model until the desired number of models is reached in this analysis. The details are presented in **Table 3**.

3.3 Best model selection

All models were statistically significant ($p < 0.001$) except models M_7 and M_{12}. Based on the Delong test, we excluded two variables (FP media exposure and wealth status) from our final analysis. The remaining significant variables were considered risk factors for predicting contraceptive practice among women aged 15–49 years in Bangladesh. From **Table 4**, Model M_{21} was the final model for analysis, and selected risk factors were also used for the final analysis. The details of the best model selection procedure are given in **Table 4**.

3.4 Performance parameter of machine learning algorithms

This study used seven different machine algorithms to classify contraceptive practices among married women both training and an experimental/test dataset.

Model	Model
M_1 = respondent age	M_{12} = M_{11} + wealth status
M_2 = M_1 + division	M_{13} = M_{12} + women education
M_3 = M_2 + religion	M_{14} = M_{13} + husband education
M_4 = M_3 + Sex of household head	M_{15} = M_{14} + sexually transmitted infection (STI)
M_5 = M_4 + residence	M_{16} = M_{15} + children ever born
M_6 = M_5 + respondent working status	M_{17} = M_{16} + number of living children
M_7 = M_6 + FP media exposure	M_{18} = M_{17} + ideal number of children
M_8 = M_7 + age at first marriage	M_{19} = M_{18} + fertility preference
M_9 = M_8 + currently breastfeeding	M_{20} = M_{19} + marital status
M_{10} = M_9 + currently amenorrhoeic	M_{21} = M_{20} + decision making for using contraception
M_{11} = M_{10} + currently abstaining	

Table 3.
Create a model-based hierarchical approach.

Model	AUC	DeLong's test for AUC (*p*-value)	Decision	Model selection
M_1	0.629	−9.26 (0.000)	M_2 has a significantly different AUC from M_1	M_2 is selected
M_2	0.660			
M_3	0.662	−2.16 (0.031)	M_3 significantly different AUC from M_2	M_3 is selected
M_4	0.713	−16.21 (0.000)	M_4 significantly different AUC from M_3	M_4 is selected
M_5	0.714	−2.03 (0.041)	M_5 had significantly different AUC from M_4	M_5 is selected
M_6	0.715	−2.24 (0.025)	M_6 had significantly different AUC from M_5	M_6 is selected
M_7	0.716	−0.61 (0.545)	M_7 had not significantly different AUC from M_6	M_7 is not selected
M_8	0.716	−2.63 (0.008)	M_8 had a significantly different AUC from M_6	M_8 is selected
M_9	0.723	−4.34 (0.000)	M_9 had a significantly different AUC from M_8	M_9 is selected
M_{10}	0.762	−14.38 (0.000)	M_{10} had a significantly different AUC from M_9	M_{10} is selected
M_{11}	0.773	−8.05 (0.000)	M_{11} had a significantly different AUC from M_{10}	M_{11} is selected
M_{12}	0.773	−0.72 (0.472)	M_{12} had not a significantly different AUC from M_{11}	M_{12} is not selected
M_{13}	0.774	−2.22 (0.029)	M_{13} had a significantly different AUC from M_{11}	M_{13} is selected
M_{14}	0.775	−2.17 (0.030)	M_{14} had a significantly different AUC from M_{13}	M_{14} is selected
M_{15}	0.776	−2.13 (0.033)	M_{15} had a significantly different AUC from M_{14}	M_{15} is selected
M_{16}	0.799	−11.81 (0.000)	M_{16} had a significantly different AUC from M_{15}	M_{16} is selected
M_{17}	0.813	−9.26 (0.000)	M_{17} had a significantly different AUC from M_{16}	M_{17} is selected
M_{18}	0.816	−4.74 (0.000)	M_{18} had a significantly different AUC from M_{17}	M_{18} is selected
M_{19}	0.828	−11.45 (0.000)	M_{19} had a significantly different AUC from M_{18}	M_{19} is selected
M_{20}	0.847	−14.69 (0.000)	M_{20} had a significantly different AUC from M_{19}	M_{20} is selected
M_{21}	0.866	−21.75 (0.000)	M_{21} had a significantly different AUC from M_{20}	M_{21} is selected

Table 4.
Best model selection based on Delong's test.

Performance parameters (such as accuracy, precision, recall, F1, specificity, and AUC value) were used to compare the predictive performance of these algorithms. In addition, Cohen Kappa's statistical information was used to determine the discriminant accuracy of the algorithm. The prediction results with performance parameters for each algorithm are shown in **Table 5** and **Figure 3**.

Table 5 shows that the logistic regression classifier has an accuracy of 78.52%. The precision and recall of the fitted model were 81.23% and 82.39%, respectively, while the F1 score was 81.81%. The area under the curve (AUC) was calculated to be 86.57%. The prediction performance result of a random forest was displayed with an accuracy of 77.57%. Here, the precision, recall, and F1 score of the random forest classifier were 73.82%, 85.35%, and 81.99%, respectively. The AUC, in this case, was 84.07%. The final accuracy of the naïve Bayes classifier was 76.56%, with a precision of 75.73% and a recall of 88.32%. The F1 score and the AUC value, in this case, were 81.54% and 84.17%, respectively. Using Least Absolute Shrinkage and Selection Operator (LASSO) analysis, the accuracy in the test data set was seen as 79.08% with precession and recall of 79.39% and 86.85% respectively, and the F1 score was 82.96%. According

Model name	Accuracy (95% CI)	Cohen's kappa	Precession	Recall	F1	AUC	Specificity
LR	78.52 (77.39, 79.61)	0.5559	81.23	82.39	81.81	86.57	73.03
RF	77.57 (76.43, 78.68)	0.5288	78.32	85.35	81.69	84.07	66.53
NB	76.56 (75.40, 77.69)	0.4995	75.73	88.32	81.54	84.17	59.90
LASSO	79.08 (77.96, 80.16)	0.5601	79.39	86.85	82.96	86.59	68.06
CT	78.57 (77.45, 79.67)	0.5464	78.16	88.06	82.81	85.59	65.13
AdaBoost	78.50 (77.37, 79.59)	0.5523	80.20	84.08	82.10	86.15	70.59
NN	79.34 (78.23, 80.42)	0.5626	78.71	88.76	83.44	86.90	65.99

Table 5.
Performance evaluation for seven ML algorithms (test data set).

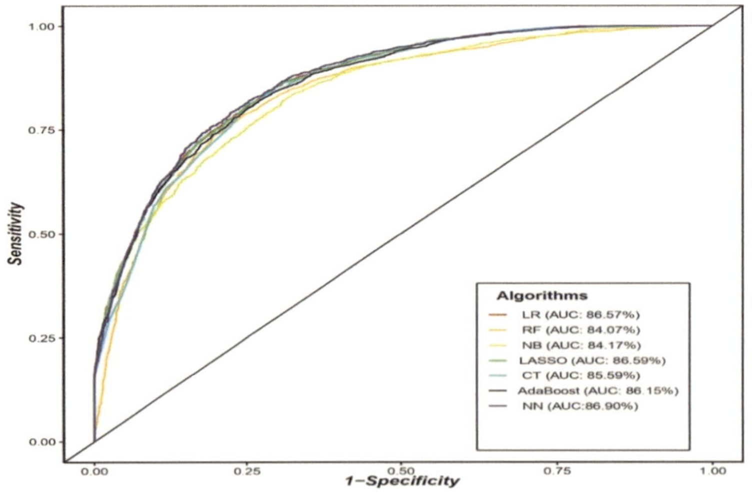

Figure 3.
Area under curve of all seven machine learning classifiers.

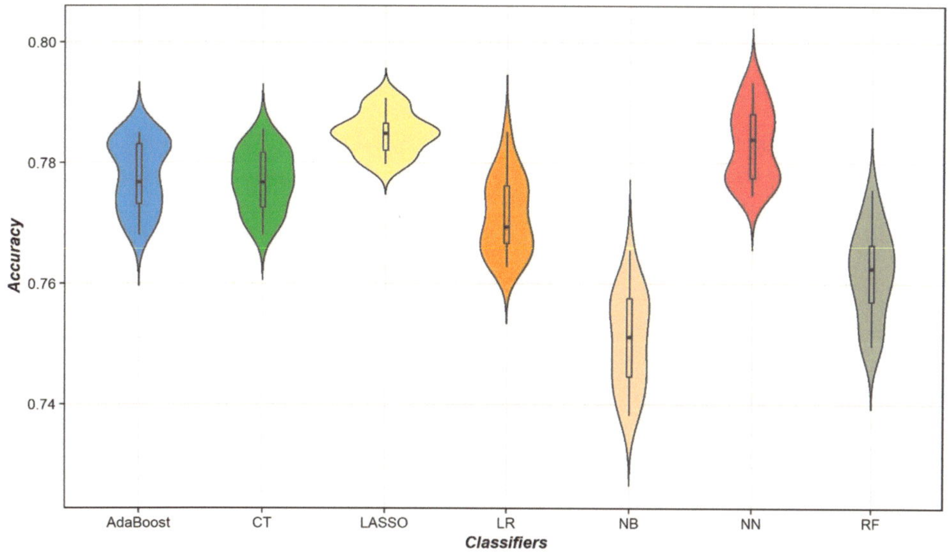

Figure 4.
Violin plots of the 10-fold cross-validation.

to the test observation results, the classification tree method showed 78.57% accuracy in predicting contraceptive practice among married women, with a precession of 78.16%, a recall of 88.06%, an F1 score of 82.81%, and an AUC value of 85.59%. For AdaBoost, these values are 78.05% (accuracy), 80.20% (precession), 84.88% (recall), 82.10% (F1 score) and 86.15% (AUC). Finally, we used an artificial neural network and obtained an accuracy of 79.34%. Here, other parameters such as precession, recall, F1 score, and AUC are 78.71%, 88.76%, 83.44%, and 86.90% respectively. Among the seven classifiers, we obtained the best performance from NN in terms of both accuracy and AUC. Cohen's kappa value is 0.5626.

This violin plot shows the relationship of seven classifiers to accuracy. The shaded areas detail the distribution of the data in each classifier. **Figure 4** shows that NN provided the highest mean accuracy, followed by LASSO and AdaBoost. Unlike the boxplot, the entire distribution of the 10-fold accuracy can be visualized in this violin plot (**Figure 4**).

4. Discussion

This is the very first study that uses a hierarchical logistic classifier in a machine learning approach. Then the predictive performance of the hierarchical logistic classifier was compared with the other six machine learning algorithms' predictive power. In this study, the use of contraception among ever-married women in Bangladesh has been predicted using sociodemographic factors. This study can provide policymakers and academics with a starting point to examine key outlines in a larger framework and raise noteworthy interventions.

The study found that the prevalence of contraception was almost 59% in Bangladesh. The prevalence rate of contraceptives in India is 54%, while the rates were 47%, 34%, and 65%, respectively, for Nepal, Pakistan, and Sri Lanka [37, 38].

As the government of Bangladesh is committed to the London Summit on Family Planning to improve contraceptive access and use among impoverished people in both urban and rural areas [39], the findings of this study will provide grounding direction for the increase in the prevalence of contraception.

In this study, we used hierarchical LR, RF, NB, LASSO, CT, AdaBoost, and NN machine learning techniques to predict contraceptive practice among ever-married women in Bangladesh. The current analysis was to evaluate which performed better based on the accurate prediction rate of contraceptive use for 2014, BDHS data sets. Moreover, there was no evidence of scientific study that used a hierarchical logistic classifier and several supervised learning. In this study, 70% of the respondents were used for model tuning purposes, and the remaining 30% were used to check model performance, for the model tuning was performed using 10-fold cross-validation on the training dataset. The researcher observed that cross-validation is most commonly used to evaluate model performance [40]. The prediction of contraceptive use was measured by performance parameters (such as accuracy, precision, recall, F1, and AUC value) compared to the performance of seven different machine learning classifiers in this analysis. Cohen's kappa, the proportion of predicted to actual classification in the dataset, is used to assimilate model perfection. Among the used models, the Neural Network outperformed other models with an accuracy of 79.34%. Additionally, in terms of Cohen kappa, the result of this analysis also highlighted that the Neural Network provides the best predictive performance (Cohen's κ = 0.5626). This indicates Neural Networks have achieved better performance than other LR, RF, Lasso Regression, NB, CT, and AdaBoost. Hailemariam et al. proposed a J48 decision tree that performed better than Naïve Bayes to predict contraceptive practice in Ethiopian women [41]. However, Hailemariam et al. have not used the neural network in their study [41]. In a data mining study in India, the CART model produces pretty satisfactory results for finding the predictors of contraception use among married women [42]. However, Vaz and his team member also found that the Random Forest model was the most accurate model for predicting women's fertile periods [43]. Machine learning algorithms can be quite helpful in predicting infertility in women, according to a study conducted in Nigeria [44].

5. Conclusions

In this paper, we investigate the hierarchical logistic regression classifier in machine learning approaches to identify potential risk factors related to contraceptive practices of women in Bangladesh. In summary, we conclude that all of the selected covariates were significant determinants for contraceptive practice except FP Mass media exposure and wealth status according to the hierarchical logistic regression classifier in machine learning approaches based on the Delong test. Here, we compared seven supervised machine learning algorithms to predict contraceptive practice among ever-married women aged between 15 and 49 years in Bangladesh. The NN model has exhibited the best results based on the performance parameters, having demonstrated an accuracy of 79.34%, a precision of 78.71%, a recall of 88.76%, an F1 score of 83.44%, and an AUC value of 86.90. Among the seven algorithms, the NN model performs the best in terms of accuracy, Cohen's kappa statistic, and area under the curve (AUC). This study recommends the use of the NN model and policymakers should pay attention to continuing this study in the future.

Acknowledgements

A special thank goes to the Demographic Health Surveys for enabling us to use Bangladesh Demographic Health Survey data for our study from https://dhsprogram.com/data/.

Funding

This study did not receive funding.

Conflicts of interest

The authors declare that they are not competing of interest.

Data availability

This study was analyzed using secondary data, which were available at "https://dhsprogram.com/data/".

Author details

Iqramul Haq[1*], Md. Ismail Hossain[2], Md. Moshiur Rahman[3], Md. Injamul Haq Methun[4], Ashis Talukder[5], Md. Jakaria Habib[2] and Md. Sanwar Hossain[2]

1 Department of Agricultural Statistics, Sher-e-Bangla Agricultural University, Dhaka, Bangladesh

2 Department of Statistics, Jagannath University, Dhaka, Bangladesh,

3 Department of Pharmacology and Toxicology, Sher-e-Bangla Agricultural University, Dhaka, Bangladesh

4 Department of Statistics, Tejgaon College, Dhaka, Bangladesh

5 Statistics Discipline, Khulna University, Khulna, Bangladesh

*Address all correspondence to: iqramul.haq@sau.edu.bd

References

[1] United Nations Population Fund. Sexual and Reproductive Health for all: Reducing Poverty, Advancing Development and Protecting Human Rights. New York, New York, United States: United Nations Population Fund; 2010

[2] United Nations. Transforming our World: The 2030 Agenda for Sustainable Development United Nations. 2015. Available from: https://sustainabledevelopment.un.org/content/documents/21252030%20Agenda%20for%20Sustainable%20Development%20web.pdf

[3] World Health Organization. Health-Related Millennium Development Goals. 2015 . Available from: https://www.who.int/gho/publications/world_health_statistics/EN_WHS2015_Part1.pdf?ua=1

[4] Cleland J, Conde-Agudelo A, Peterson H, Ross J, Tsui A. Contraception and health. The Lancet. 2012;**380**(9837):149-156. DOI: 10.1016/s0140-6736(12)60609-6

[5] Ahmed S, Li Q, Liu L, Tsui AO. Maternal deaths averted by contraceptive use: An analysis of 172 countries. The Lancet. 2012;**80**(9837): 111-125. DOI: 10.1016/S0140-6736(12) 60478-4

[6] Brunner Huber LR, Smith K, Sha W, Vick T. Interbirth interval and pregnancy complications and outcomes: Findings from the pregnancy risk assessment monitoring system. Journal of Midwifery & Women's Health. 2018; **63**(4):436-445. DOI: 10.1111/jmwh.12745

[7] Darroch J. Singh S. Estimating Unintended Pregnancies Averted from Couple-Years of Protection (CYP). 2011. Available from: https://www.guttmacher.org/sites/default/files/page_files/guttmacher-cyp-memo.pdf

[8] Liu L, Becker S, Tsui A, Ahmed S. Three methods of estimating births averted nationally by contraception. Population Studies. 2008;**62**(2):191-210. DOI: 10.1080/00324720801897796

[9] Yazdkhasti M, Pourreza A, Pirak A, Abdi F. Unintended pregnancy and its adverse social and economic consequences on health system: A narrative review article. Iranian Journal of Public Health. 2015;**44**(1):12-21

[10] Aviisah PA, Dery S, Atsu BK, Yawson A, Alotaibi RM, Rezk HR, et al. Modern contraceptive use among women of reproductive age in Ghana: Analysis of the 2003–2014 Ghana demographic and health surveys. BMC Women's Health. 2018;**18**(1):1-10. DOI: 10.1186/s12905-018-0634-9

[11] Haq I, Sakib S, Talukder A. Sociodemographic factors on contraceptive use among ever-married women of reproductive age: Evidence from three demographic and health surveys in Bangladesh. Medical Science. 2017;**5**(4):31. DOI: 10.3390/medsci5040031

[12] Kopp Kallner H, Thunell L, Brynhildsen J, Lindeberg M, Gemzell Danielsson K. Use of contraception and attitudes towards contraceptive use in Swedish women—A Nationwide survey. PLoS One. 2015;**10**(5):e0125990. DOI: 10.1371/journal.pone.0125990

[13] Mandiwa C, Namondwe B, Makwinja A, Zamawe C. Factors associated with contraceptive use among young women in Malawi: Analysis of the 2015–16 Malawi demographic and health survey data. Contraception and

Reproductive Medicine. 2018;**3**(1):12-19. DOI: 10.1186/s40834-018-0065-x

[14] Moazenzadeh R, Mohammadi B, Shamshirband S, Chau K. Coupling a firefly algorithm with support vector regression to predict evaporation in northern Iran. Engineering Applications of Computational Fluid Mechanics. 2018;**12**(1):584-597. DOI: 10.1080/19942060.2018.1482476

[15] Mousa SR, Bakhit PR, Osman OA, Ishak S. A comparative analysis of tree-based ensemble methods for detecting imminent lane change maneuvers in connected vehicle environments. Transportation Research Record: Journal of the Transportation Research Board. 2018;**2672**(42):268-279. DOI: 10.1177/0361198118780204

[16] Zhang Y, Haghani A. A gradient boosting method to improve travel time prediction. Transportation Research Part C: Emerging Technologies. 2015;**58**:308-324. DOI: 10.1016/j.trc.2015.02.019

[17] NIPORT, Mitra and Associates, & ICF International. Bangladesh Demographic and Health Survey 2014. Bangladesh: NIPORT, Mitra and Associates, and ICF International; 2016

[18] Johnson EO. Determinants of modern contraceptive uptake among Nigerian women: Evidence from the National Demographic and health survey. African Journal of Reproductive Health. 2017;**21**(3):89-95. DOI: 10.29063/ajrh2017/v21i3.8

[19] Gebre MN, Edossa ZK. Modern contraceptive utilization and associated factors among reproductive-age women in Ethiopia: Evidence from 2016 Ethiopia demographic and health survey. BMC Women's Health. 2020;**20**(1):1-14. DOI: 10.1186/s12905-020-00923-9

[20] Islam AZ, Mondal MNI, Khatun ML, Rahman MM, Islam MR, Mostofa MG, et al. Prevalence and determinants of contraceptive use among employed and unemployed women in Bangladesh. International Journal of MCH and AIDS. 2016;**5**(2):92-102. DOI: 10.21106/ijma.83

[21] Kidayi PL, Msuya S, Todd J, Mtuya CC, Mtuy T, Mahande MJ. Determinants of modern contraceptive use among women of reproductive age in Tanzania: Evidence from Tanzania demographic and health survey data. Advances in Sexual Medicine. 2015;**05**(03):43-52. DOI: 10.4236/asm.2015.53006

[22] Solanke BL. Factors influencing contraceptive use and non-use among women of advanced reproductive age in Nigeria. Journal of Health, Population and Nutrition. 2017;**36**(1):1-14. DOI: 10.1186/s41043-016-0077-6

[23] Sridhar A, Salcedo J. Optimizing maternal and neonatal outcomes with postpartum contraception: Impact on breastfeeding and birth spacing. Maternal Health, Neonatology and Perinatology. 2017;**3**(1):1-10. DOI: 10.1186/s40748-016-0040-y

[24] Vu LTH, Oh J, Bui QT-T, Le AT-K. Use of modern contraceptives among married women in Vietnam: A multilevel analysis using the multiple indicator cluster survey (2011) and the Vietnam population and housing census (2009). Global Health Action. 2016;**9**(1):29574. DOI: 10.3402/gha.v9.29574

[25] Zhang L, Zhang B. Hierarchical machine learning–a learning methodology inspired by human intelligence. In: International Conference on Rough Sets and Knowledge Technology. Berlin, Heidelberg: Springer; 2006. pp. 28-30. DOI: 10.1007/11795131_3

[26] DeLong ER, DeLong DM, Clarke-Pearson DL. Comparing the areas under two or more correlated receiver operating characteristic curves: A nonparametric approach. Biometrics. 1988;**44**(3):837. DOI: 10.2307/2531595

[27] Anuse A, Vyas V. A novel training algorithm for convolutional neural network. Complex & Intelligent Systems. 2016;**2**(3):221-234. DOI: 10.1007/s40747-016-0024-6

[28] Buntine W. Learning classification trees. Statistics and Computing. 1992; **2**(2):63-73. DOI: 10.1007/bf01889584

[29] Jang W, Lee JK, Lee J, Han SH. Naïve Bayesian classifier for selecting good/bad projects during the early stage of international construction bidding decisions. Mathematical Problems in Engineering. 2015;**2015**:1-12. DOI: 10.1155/2015/830781

[30] Talukder A, Ahammed B. Machine learning algorithms for predicting malnutrition among under-five children in Bangladesh. Nutrition. 2020;**78**: 110861. DOI: 10.1016/j.nut.2020.110861

[31] Vasquez MM, Hu C, Roe DJ, Chen Z, Halonen M, Guerra S. Least absolute shrinkage and selection operator type methods for the identification of serum biomarkers of overweight and obesity: Simulation and application. BMC Medical Research Methodology. 2016;**16**(1): 154-172. DOI: 10.1186/s12874-016-0254-8

[32] Wu P, Zhao H. Some analysis and research of the AdaBoost algorithm. In: International Conference on Intelligent Computing and Information Science. Berlin, Heidelberg: Springer; 2011. pp. 1-5

[33] Xu X, Xia L, Zhang Q, Wu S, Wu M, Liu H. The ability of different imputation methods for missing values in mental measurement questionnaires. BMC Medical Research Methodology. 2020;**20**(1):1-9. DOI: 10.1186/s12874-020-00932-0

[34] Madley-Dowd P, Hughes R, Tilling K, Heron J. The proportion of missing data should not be used to guide decisions on multiple imputation. Journal of Clinical Epidemiology. 2019; **110**:63-73. DOI: 10.1016/j.jclinepi.2019.02.016

[35] Liu B, Fang L, Liu F, Wang X, Chou K-C. iMiRNA-PseDPC: microRNA precursor identification with a pseudo distance-pair composition approach. Journal of Biomolecular Structure & Dynamics. 2016;**34**(1):223-235. DOI: 10.1080/07391102.2015.1014422

[36] Liaw A, Wiener M. Classification and regression by random Forest. R News. 2002;**2**(3):18-22. Available from: https://cogns.northwestern.edu/cbmg/LiawAndWiener2002.pdf

[37] Family Planning. India: Commitment Maker since 2012. 2018. Available from: https://www.familyplanning2020.org/india

[38] The World Bank. Contraceptive Prevalence, Any Methods (% of Women Ages 15–49) Data. 2019. Available from: https://data.worldbank.org/indicator/SP.DYN.CONU.ZS%20(2019)

[39] Huda FA, Robertson Y, Chowdhuri S, Sarker BK, Reichenbach L, Somrongthong R. Contraceptive practices among married women of reproductive age in Bangladesh: A review of the evidence. Reproductive Health. 2017;**14**(1):69-77. DOI: 10.1186/s12978-017-0333-2

[40] Cawley GC, Talbot NLC. Gene selection in cancer classification using sparse logistic regression with Bayesian

regularization. Bioinformatics. 2006; **22**(19):2348-2355. DOI: 10.1093/ bioinformatics/btl386

[41] Hailemariam T, Gebregiorgis A, Meshesha M, Mekonnen W. Application of data mining to predict the likelihood of contraceptive method use among women aged 15-49 case of 2005 demographic health survey data collected by central statistics agency, Addis Ababa, Ethiopia. Journal of Health & Medical Informatics. 2017;**8**(3): 274-279. DOI: 10.4172/2157-7420. 1000274

[42] Chaurasia AR. Contraceptive use in India: A data mining approach. International Journal of Population Research. 2014;**2014**:1-11. DOI: 10.1155/ 2014/821436

[43] Vaz F, Silva RR, Bernardino J. Using data mining in a mobile application for the calculation of the female fertile period. In: Proceedings of the 10th International Joint Conference on Knowledge Discovery, Knowledge Engineering and Knowledge Management. Setúbal, Portugal: SciTePress; 2018. DOI: 10.5220/ 0007228603590366

[44] Balogun JA, Egejuru N, Idowu P. Comparative analysis of predictive models for the likelihood of infertility in women using supervised machine learning techniques. Computer Reviews Journal. 2018;**2**(1):313-330

Chapter 2

Evaluating Similarities and Differences between Machine Learning and Traditional Statistical Modeling in Healthcare Analytics

Michele Bennett, Ewa J. Kleczyk, Karin Hayes and Rajesh Mehta

Abstract

Data scientists and statisticians are often at odds when determining the best approaches and choosing between machine learning and statistical modeling to solve their analytical challenges and problem statements across industries. However, machine learning and statistical modeling are actually more closely related to each other rather than being on different sides of an analysis battleground. The decision on which approach to choose is often based on the problem at hand, expected outcome(s), real world application of the results and insights, as well as the availability and granularity of data for the analysis. Overall machine learning and statistical modeling are complementary techniques that are guided on similar mathematical principles, but leverage different tools to arrive at insights. Determining the best approach should consider the problem to be solved, empirical evidence and resulting hypothesis, data sources and their completeness, number of variables/data elements, assumptions, and expected outcomes such as the need for predictions or causality and reasoning. Experienced analysts and data scientists are often well versed in both types of approaches and their applications, hence use best suited tools for their analytical challenges. Due to the importance and relevance of the subject in the current analytics environment, this chapter will present an overview of each approach as well as outline their similarities and differences to provide the needed understanding when selecting the proper technique for problems at hand. Furthermore, the chapter will also provide examples of applications in the healthcare industry and outline how to decide which approach is best when analyzing healthcare data. Understanding of the best suited methodologies can help the healthcare industry to develop and apply advanced analytical tools to speed up the diagnostic and treatment processes as well as improve the quality of life for their patients.

Keywords: machine learning, statistical modeling, data science, healthcare analytics, research design

1. Introduction

In the recent years, machine learning techniques have been utilized to solve problems at hand across multitudes of industries and topics. In the healthcare industry, these techniques are often applied to a variety of healthcare claims and electronic health records data to garner valuable insights into diagnostic and treatment pathways in order to help optimize patient healthcare access and treatment process [1]. Unfortunately, many of these applications resulted in inaccurate or irrelevant research results, as proper research protocols were not fully followed [2]. On the other hand, statistics has been the basis of analysis in healthcare research for decades, especially, in the areas of clinical trials and health economics and outcomes research (HEOR), where the precision and accuracy of analyses have been the primary objectives [3]. Furthermore, the classical statistics methodologies are often preferred in those research areas to ensure the ability to replicate and defend the results and ultimately, the ability to publish the research content in peer-reviewed medical journals [3]. The increased availability of data, including data from wearables, provided the opportunity to apply a variety of analytical techniques and methodologies to identify patterns, often hidden, that could help with optimization of healthcare access as well as diagnostic and treatment process [4].

With the rapid increase in data from the healthcare and many other industries, it is important to consider how to select well - suited statistical and machine learning methodologies that would be best for the problem at hand, the available data type, and the overall research objectives [5]. Machine learning alone or complemented by statistical modeling is becoming, not just a more common, but a desired convergence to take advantage of the best of both approaches for advancing healthcare outcomes [1]. Please note that this book chapter was originally posted on the Cornell University's research working article website: https://arxiv.org. The book chapter content is mostly the same between the two versions [6].

2. Machine learning foundation is in statistical learning theory

Machine learning (ML) is considered a branch of artificial intelligence and computer science that focuses on mimicking human behaviors through a set of algorithms and methods that use historical values to predict new values [7], without specifically being coded to do so and thereby learning over time [8, 9]. ML is grounded in statistical learning theory (SLT), which provides the constructs used to create prediction functions from data. One of the first examples of SLT was the creation of the support vector machine (SVM), the supervised learning method that can be used as for both classification and regression and has become a standard in modeling how to recognize visual objects [7]. SLT formalizes the model that makes a prediction based on observations (i.e., data) and ML automates the modeling [7].

SLT sets the mathematical and theoretical framework for ML as well as the properties of learning algorithms [7] with the goals of providing mechanisms for studying inference and creating algorithms that become more precise and improved over time [8]. SLT is based multivariate statistics and functional analysis [8]. Functional analysis is the branch of statistics that measures shapes, curves, and surfaces, extending multivariate vector statistics to continuous functions and finding functions that describe data patterns [8]. Inductive inference is the process of generalizing and modeling past observations to make predictions for the future; SLT formalizes the modeling concepts of inductive inference, while ML automates them [8].

For example, pattern recognition is considered a problem of inductive inference and SLT, as it is a curving-fitting problem, and one of the most common applications of ML [7–9]. Pattern recognition is not suited for traditional computer programming as the inferences needed are not free of assumptions and the patterns are not easily described or labeled programmatically with deterministic functions. The standard mathematics behind SLT makes no assumptions on distributions, uses stochastic functions that can include humans labeling the "right" classification, i.e., training data, and can assume that the probability of the occurrence of one observation is independent of another thereby including the concept of randomness [7–9]. These tenets are therefore those of ML as well.

SLT also provides the definition of terms often using in ML such as overfitting, underfitting and generalization. Overfitting is when the presence of noise in the data negatively affects training and the ultimate model performance because the noise is being incorporated into the learning process, thereby giving error when the model sees new data [8, 9]. Underfitting is when the noise impacts both performance on training data as well as new and unseen data [9]. In ML, discussion about underfitting and overfitting are often used to describe models that do not generalize the data effectively and might not present the right set of data elements to explain the data patterns and posited hypotheses [9]. Underfitting is often defined when model which is missing features that would be present in the most optimized model, akin to a regression model not fully explaining all of the variance of the dependent variable [9]. In a similar vein, overfitting is when the model contains more features or different features than is optimal, like a regression model with autocorrelation or multicollinearity [9].

The general goal of learning algorithms and therefore ML model optimization is to reduce the dimensions, features, or data variables to the fewest number needed as that reduces noise or the impact of trivial variables that can overfit or unfit [8, 9]. A regularization model can then become generalized to perform not just on the past or the training data, but also on future and yet unseen data [8, 9]. Although true generalization needs both the right modeling criteria as well as strong subject matter knowledge [8].

Often dimension reduction approaches like Principal Component Analysis (PCA) or boot strapping techniques used along with subject matter expertise can help resolve how to refine models, combat fit challenges, as well as improve generalization potential [9, 10]. Furthermore, understanding the studied population and data characteristics can further help define the data to be used, variable selection, and proper model set up [10].

3. Similarities between machine learning and statistical modeling

Statistical modeling is based on SLT and use of mathematical models and statistical assumptions to generate sample data and make predictions about the real world occurrences. A statistical model is often represented as a collection of probability distributions on a set of all possible outcomes. Furthermore, statistical modeling has evolved in the last few decades and shaped the future of business analytics and data science, including the current use and applications of ML algorithms. On the other hand, machine learning does not require many assumptions and interventions when running algorithms in order to accurately predict studied outcomes [7].

There are similarities between ML and statistical modeling that are prevalent across most analytical efforts. Both techniques use historical data as input to predict new output values, but they vary as noted above on the underlying assumptions and the level of analyst intervention and data preparation.

Overall, machine learning foundations are based from statistical learning theory, and it is recommended for the data scientists to apply SLT's guiding rules during analysis. While it may seem as a statistical background and understanding is not required when analyzing the underlying data, this misconception often leads to data scientist's inability to set up proper research hypothesis and analysis due to a lack of understanding of the problem and the underlying data assumptions as well as caveats. This issue can in turn result in biased and irrelevant results as well as unfounded conclusions and insights. With that in mind, it is important to evaluate the problem at hand, and consider both statistical modeling and ML as possible methods to be applied. Understanding the underlying assumptions of the data and statistical inference can help support proper technique selection and guide the pathway to solution [11]. In the later sections of the chapter, application of both techniques will be provided and the reasoning for selecting the methods presented to guide future research.

As mentioned above, the similarities between ML and statistical modeling start with the underlying assumption that data or observations from the past can be used to predict the future [7]. The variables included in the analysis generally represent two types: dependent variables, that in ML are called targets, and independent variables, that in ML are called features. The definition of the variables is the same across both techniques [8]. Furthermore, both ML and statistical modeling leverage the available data in a way that allow for generalization of results to larger population [7]. The loss and risk associated with the models accuracy and representation of the real world occurrence is described frequently in terms of mean squared error (MSE). In statistical modeling, MSE is the difference between the predicted value and the actual value and is used to measure loss of the performance of predictions. In the ML, the same MSE concept is presented via a confusion matrix that evaluates a classification problem's accuracy [9].

4. Differences between machine learning and statistical modeling

Differences between machine learning and statistical modeling are distinct and based on purposes and needs for the analysis as well as the outcomes. Assumptions and purposes for the analysis and approach can vastly differ. For example, statistics typically assumes that predictors or features are known and additive, models are parametric, and testing of hypotheses and uncertainty are forefront. On the other hand, ML does not make these assumptions [12]. In ML, many models are based on non-parametric approaches where the structure of model is not specified or unknown, additivity is not expected, and assumptions about normal distributions, linearity or residuals, for example, are not needed for modeling [10].

The purpose of ML is predictive performance using general purpose learning algorithms to find patterns that are less known, unrelated, and in complex data without a priori view of underlying structures [10]. Whereas in statistical modeling, consideration for inferences, correlations, and the effects of a small number of variables are drivers [12].

Due to the differences in the methods' characteristics, it is important to understand the variations in application of the techniques when solving healthcare problems. For example, one typical application of statistics is to analyze whether a population has a particular medical condition. For some diseases such as diabetes, the condition is easily screened for and diagnosed using distinct lab values, such as elevated and increasing HbA1C over time, high glucose levels and low insulin levels, often due to insulin depletion occurring from unmanaged diabetes. Also conditions such as hypertension can easily be detected at home or in the healthcare provider's office using simple blood pressure measurement and monitoring, and wearables can identify when patients are experiencing atrial fibrillation, abnormal heart rhythms and even increased patient falls (possible syncope). Therefore, analyses of patients with these easily measurable conditions can be done simply by qualifying patients based on lab values or biomarkers falling within or outside of certain ranges. One of the simplest examples is identifying patients with diabetes [13]. This can be accomplished by using A1C levels to group patients as having no diabetes (A1C < 5.7), pre-diabetes (AIC of 5.7–6.4), or diabetes (A1C > 6.4). These ranges are based on American Diabetes Association Diagnosis Guidelines and a very high, medically accepted correlation between AIC levels and the diagnosis of diabetes [14].

On the other hand, if the objective of the research is to predict which pre-diabetic patients are most likely to progress to diabetes, a myriad of factors influence diabetes progression including extent of chronic kidney disease, high blood pressure, insulin levels over time, body mass index/obesity, age, years with diabetes, success of prior therapy, number and types of prior therapies, family history, coronary artery disease, prior cardiovascular events, infections, etc. A complicated combination of comorbidities, risk factors, and patient behavior can lead to differing diabetes complications and varying outcomes makes prediction more challenging and thus it represents a good candidate for the use of machine learning techniques. Classification models such as gradient boosting tree algorithms have been used to successfully predict diabetes progression, especially earlier in the disease. While there any many diabetes risk factors and co-morbidities, these disease characteristics are well studied over many years, thus enabling stable predictive models which perform well over time [14].

Overall, machine learning is highly effective when the model uses more than a handful of independent variables/features [10]. ML is required when the number of features (p) is larger than the number of records or observations (n) – this is called the curse of dimensionality [15, 16], which increases the risk of overfitting, but can be overcome with dimensionality reductive techniques (i.e., PCA), as part of modeling [15] and clinical/expert input on the importance or lack thereof of certain features, is it relates to the disease or its treatment. Additionally, statistical learning theory teaches that learning algorithms increase their ability to translate complex structures from data at a greater and faster rate than the increase of sample size capture can alone provide [8]. Therefore, statistical learning theory and ML offer methods for addressing high-dimensional data or big data (high velocity, volume and variety) and smaller sample sizes [17], such as recursive feature elimination and support vector machines, boosting, or cross validation which can also minimize prediction error [18].

In the healthcare industry, machine learning models are frequently used in cancer prediction, generally in three areas: (1) predicting a patient with a cancer prognosis/diagnosis, (2) predicting cancer progression, and (3) predicting cancer mortality. Of these, predicting whether a patient may have a cancer prognosis/diagnosis can be more or less difficult depending on the tumor type. Certain cancers such as lung cancer, breast cancer, prostate cancer, and skin cancer are evaluated based on specific signs and symptoms, and non-invasive imaging or blood tests. These cancers are easier to predict. Conversely, cancers with non-descript symptoms such fatigue, dizziness, GI pain and distress, and lack of appetite are much more difficult to predict even with machine learning models as these symptoms are associated with multiple tumor types (for example esophageal, stomach, bladder, liver, and pancreatic cancer) and also mimic numerous other conditions [14].

For cancers with vague symptoms, understanding the patient journey is very important to cancer prediction. If a prediction period is too long and does not reflect the time period before diagnosis when symptoms develop, the model may overfit due to spurious variables not related to the condition. If the prediction period is too short, key risk factors from the patient record could be missing. Variable pruning is required in these situations. A multi-disciplinary team including business and clinical experts can help trim unrelated variables and improve model performance [14].

Model validation is an inherent part of the ML process where the data is split into training data and test data, with the larger portion of data used to train the model to learn outputs based on known inputs. This process allows for rapid structure knowledge for primary focus on building the ability to predict future outcomes [15]. Beyond initial validation of the model within the test data set, the model should be further tested in the real world using a large, representative, and more recent sample of data [19]. This can be accomplished by using the model to score the eligible population and using a look forward period to assess incidence or prevalence of the desired outcome. If the model is performing well, probability scores should be directly correlated to incidence/prevalence (the higher the probability score, the higher the incidence/prevalence). Model accuracy, precision, and recall can also be assessed using this approach [20].

Epidemiology studies and prior published machine learning research in related areas of healthcare can help benchmark the performance of the model relative to the baseline prevalent or incident population for the condition to be predicted. Machine learning models created using a few hundred or thousand patients often do not perform as well in the real world. Careful variable pruning, cohort refinement and adjustment of modeling periods can often resolve model performance problems. Newer software can be used to more quickly build, test, and iterate models, allowing users to easily transform and combine features as well as run many models simultaneously and visualize model performance, diagnosis and solve model issues [21].

5. How to choose between machine learning and statistical modeling

Machine learning algorithms are a preferred choice of technique vs. a statistical modeling approach under specific circumstances, data configurations, and outcomes needed.

5.1 Importance of prediction over causal relationships

As noted above, machine learning algorithms are leveraged for prediction of the outcome rather than present the inferential and causal relationship between the outcome and independent variables/data elements [17, 22]. Once a model has been created, statistical analysis can sometime elucidate and validate the importance and relationship between independent and dependent variables.

5.2 Application of wide and big dataset(s)

Machine Learning algorithms are learner algorithms and learn on large amount of data often presented by a large number of data elements, but not necessarily with many observations [23]. Ability of multiple replications of samples, cross validation or application of boot strapping techniques for machine learning allows for wide datasets with many data elements and few observations, which is extremely helpful in predicting rare disease onset [24] as long as the process is accompanied with real world testing to ensure the models are not suffering from overfitting [18, 19]. With the advent of less expensive and more powerful computing power and storage, multialgorithm, ensembled models using larger cohorts can be more efficiently built. Larger modeling samples that are more representative of the overall population can help reduce the likelihood of overfitting or underfitting [25]. A large cohort imposes various issues and of priority is the ability to identify the set of independent variables that are most meaningful and impactful. These significant independent variables provide a predictive and/or inferential model that can be readily acceptable in providing a real-world application. The variables in such instances may also result into more realistic magnitude and direction of the causal relationship between the independent and outcomes variables of interest.

A recent example for a real-world example in healthcare for machine learning algorithm application is to identify the likelihood of hospitalization for high-risk patients diagnosed with Covid 19. The dataset leveraged included over 20,000 independent variables across healthcare claims data for diagnostics and treatment variables. The best optimal ML model consisted of approximately 200 important predictors variables such as age, diagnosis like Type 2 diabetes/CKD/Hypertension, frequency of office visits, Obesity amongst others. None of the variables in this example were 'new', however, the magnitude and direction as a result of the ML exercise may illustrate the 'true' impact of each independent variable, a feature that is a serious limitation in traditional statistical modeling [26].

Furthermore, as explained above, statistical models tend to not operate well on very large datasets and often require manageable datasets with a fewer number of pre-defined attributes/data elements for analysis [23]. The recommended number of attributes is up to 12 in a statistical model, because these techniques are highly prone to overfitting [25]. This limitation creates a challenge when analyzing large healthcare datasets and require application of dimension reduction techniques or expert guidance in allowing to eliminate the number of independent variables in the study [23].

5.3 Limited data and model assumptions are required

In machine learning algorithms, there are fewer assumptions that need to be made on the dataset and the data elements [5]. However, a good model is usually preceded by profiling of the target and control groups and some knowledge of the domain. Understanding relationships within the data improve outcomes and interpretability [27].

Machine learning algorithms are comparatively more flexible than statistical models, as they do not require making assumptions regarding collinearity, normal distribution of residuals, etc. [5]. Thus, they have a high tolerance for uncertainty in variable performance (e.g., confidence intervals, hypothesis tests [28]. In statistical modeling emphasis is put in uncertainty estimates, furthermore, a variety of assumptions have to be satisfied before the outcome from a statistical model can be trusted and applied [28]. As a result, the statistical models have a low uncertainty tolerance [25].

Machine learning algorithms tend to be preferred over statistical modeling when the outcome to be predicted does not have a strong component of randomness, e.g., in visual pattern recognition an object must be an E or not an E [5], and when the learning algorithm can be trained on an unlimited number of exact replications [29].

ML is also appropriate when the overall prediction is the goal, with less visibility to describe the impact of any one independent variable or the relationships between variables [30], and when estimating uncertainty in forecasts or in effects of selected predictors is not a requirement [28]. However, often data scientists and data analysts leverage regression analytics to understand the estimated impact, including directionality of the relationships between the outcome and data elements, to help with model interpretation, relevance, and validity for the studied [27]. ML is also preferred when the dataset is wide and very large [23] with underlying variables are not fully known and previously described [5].

6. Machine learning extends statistics

Machine learning requires no prior assumptions about the underlying relationships between the data elements. It is generally applied to high dimensional data sets and does not require many observations to create a working model [5]. However, understanding the underlying data will support building representative modeling cohorts, deriving features relevant for the disease state and population of interest, as well as understanding how to interpret modeling results [19, 27].

In contrast, statistical model requires a deeper understanding how the data was collected, statistical properties of the estimator (p-value, unbiased estimators), the underlying distribution of the population, etc. [17]. Statistical modeling techniques are usually applied to low dimensional data sets [25].

7. Machine learning can extend the utility of statistical modeling

Robert Tibshirani, a statistician and machine learning expert at Stanford University, calls machine learning "glorified statistics," which presents the dependence of machine learning techniques on statistics in a successful execution that not only allows for a high level of prediction, but interpretation of the results to ensure validity and applicability of the results in the healthcare [17]. Understanding the association and knowing their differences enables data scientists and statisticians to expand their knowledge and apply variety of methods outside their domain of expertise. This is the notion of "data science," which aims to bridge the gap between the areas as well as bring other important to consider aspects of research [5]. Data science is evolving beyond statistics or more simple ML approaches to incorporate self-learning and autonomy with the ability to interpret context, assess and fill in data gaps, and make modeling adjustment over time [31]. While these modeling approaches are not perfect and more difficult to interpret, they provide exciting new options for difficult to solve problems, especially where the underlying data or environment is rapidly changing [27].

Collaboration and communication between not only data scientists and statisticians but also medical and clinical experts, public policy creators, epidemiologists, etc. allows for designing successful research studies that not only provide predictions and insights on relationships between the vast amount of data elements and health outcomes [30], but also allow for valid, interpretable and relevant results that can be applied with confidence to the project objectives and future deployment in the real [30, 32].

Finally, it is important to remember that machine learning foundations are based in statistical theory and learning. It may seem machine learning can be done without a sound statistical background, but this leads to not really understanding the different nuances in the data and presented results [17]. Well written machine learning code does not negate the need for an in-depth understanding of the problem, assumptions, and the importance of interpretation and validation [29].

8. Specific examples in healthcare

As mentioned earlier in the chapter, machine learning algorithms can be leveraged in the healthcare industry to help evaluate a continuum of access, diagnostic and treatment outcomes, including prediction of patient diagnoses, treatment, adverse events, side effects, and improved quality of life as well as lower mortality rates [24].

As shown in **Figure 1**, often these algorithms can be helpful in predicting a variety of disease conditions and shortening the time from awareness to diagnosis and treatment, especially in rare and underdiagnosed conditions, estimate the 'true' market size, predicting disease progression such as identifying fast vs. slow progressing patients as well as determinants of suitable next line change [32]. Finally, the models can be leveraged for patient and physician segmentation and clustering to identify appropriate targets for in-person and non-personal promotion [30].

There are, however, instances in which machine learning might not be the right tool to leverage, including when the condition or the underlying condition have a few known variables, when the market is mature and has known predetermined diagnostic and treatment algorithm, and when understanding correlations and inference is more important than making prediction [5].

One aspect of the machine learning process is to involve a cross functional team of experts in the healthcare area to ensure that the questions and problem statement along with hypothesis are properly set up [33, 34]. Many therapeutic areas require in-depth understanding of the clinical and medical concepts (i.e., diagnostic process, treatment regimens, potential adverse effects, etc.), which can help with the research design and selection of the proper analytical techniques. If the expert knowledge is not considered or properly captured in the research design, it might lead to irrelevant, invalid, and biased results, and ultimately invalidate the entire research study [33, 34].

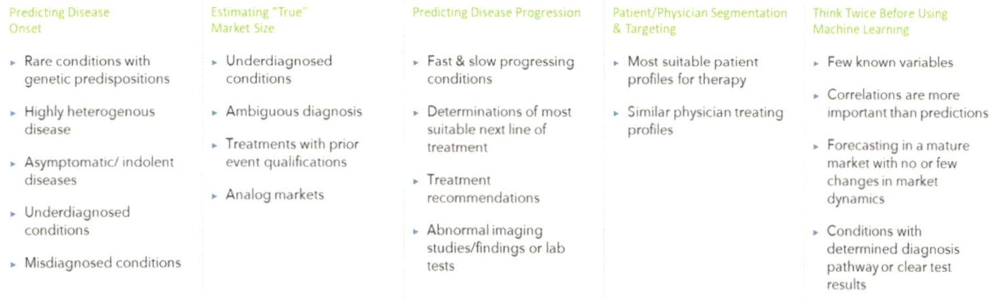

Figure 1.
Examples of Machine Learning Applications in Healthcare Analytics [22].

9. A practical guide to the predominant approach

Using a real example of a project with the goal of predicting the risk of hypertension due to underlying comorbid conditions or induced by medication, the decision to lead with machine learning vs. statistical modeling can be based on explicit criteria that can be weighed and ranked based on the desired outcome of the work [17, 32]. Please see **Figure 2** presenting an example of the approach.

As shown in **Figure 2**, pending the research objectives, machine learning or statistical modeling or both techniques could be the right method(s) to apply. For example, shifts in market trends, including shifts in patient volume of diagnosis and treatment present a suitable example when a statistical modeling type of analysis should be utilized. On the other hand, trying to predict patients with a high risk for hypertension requires the utilization of ML approaches. Leveraging both methods is best suited when predictive power and explanatory reasoning is needed to understand the important factors driving the outcome and their relative magnitudes and inferences.

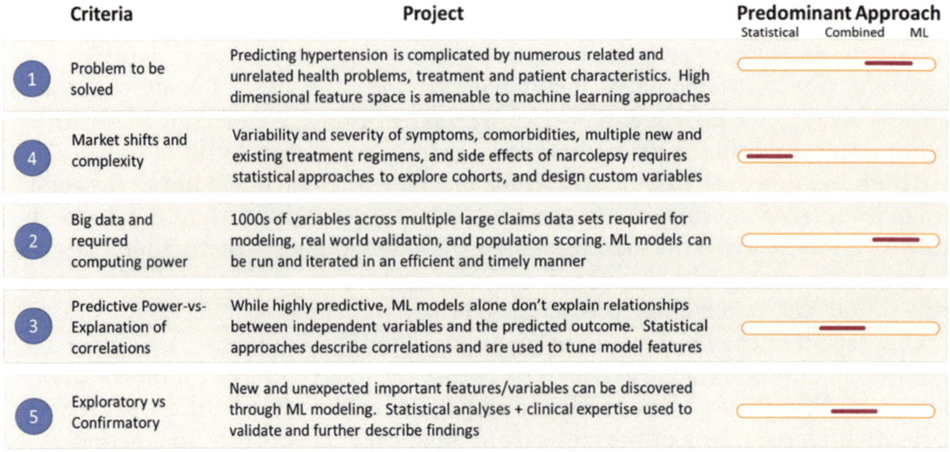

Figure 2.
Criteria for Choosing the Predominant Approach for a Project.

10. Conclusions

Machine learning requires fewer assumptions about the underlying relationships between the data elements. It is generally applied to high dimensional data sets and require fewer observations to create a working model [5]. In contrast, statistical model requires an understanding of how the data was collected, statistical properties of the estimator (p-value, unbiased estimators), the underlying distribution of the population, etc. [17]. Statistical modeling techniques are usually applied to low dimensional data sets [25]. Statistical modeling and ML are not at odds but rather complementary approaches that offer choice of techniques based on need and desired outcomes. Data scientists and analysts should not necessarily have to choose between either machine learning or statistical modeling as a mutually exclusive decision tree. Instead, selected approaches from both areas should be considered as both types of methodologies are based on the same mathematical principles but expressed somewhat differently [5, 10].

Note: This book chapter was originally posted on the Cornell University's research working paper website: https://arxiv.org. The content of the book chapter is mostly the same compared to the version posted on https://arxiv.org [6].

Funding

Authors work for Symphony Health, ICON plc Organization.

Conflict of interest

The authors declare no conflict of interest.

Author details

Michele Bennett[1,2], Ewa J. Kleczyk[1,3*], Karin Hayes[4] and Rajesh Mehta[1]

1 Symphony Health, ICON, plc Organization, Blue Bell, PA, USA

2 Data Science, Computer Science, and Business Analytics, Grand Canyon University, USA

3 The School of Economics, The University of Maine, Orono, ME, USA

4 Symphony Health, ICON, plc Organization, Phoenix, AZ, USA

*Address all correspondence to: ewa.kleczyk@symphonyhealth.com

References

[1] Beam AL, Kohane IS. Big data and machine learning in health care. JAMA. 2018;**19**(13):1317-1318. DOI: 10.1001/jama.2017.18391

[2] Shelmerdine et al. Review of study reporting guidelines for clinical studies using artificial intelligence in healthcare. BMJ Health & Care Informatics. 2021;**28**(1):e100385. DOI: 10.1136/bmjhci-2021-100385

[3] Romano R, Gambale E. Statistics and medicine: The indispensable know-how of the researcher. Translational Medicine @UniSa. 2013;**5**:28-31

[4] Razzak et al. Big data analytics for preventive medicine. Neural Computing and Application. 2020;**32**:4417-4451. DOI: 10.1007/s00521-019-04095-y

[5] Bzdok D, Altman N, Krzywiniski M. Statistics versus machine learning. Nature Methods. 2018;**15**(4):233-234. DOI: 0.1038/nmeth.4642

[6] Bennett M, Hayes K, Kleczyk EJ, Mehta R. Analytics in healthcare: Similarities and differences between machine learning and traditional advanced statistical modeling. Cornell University. 2022:1-16. Available from: https://arxiv.org/abs/2201.02469

[7] Von Luxburg U, Scholkopf B. Inductive logic. In: Handbook and History of Logic. Vol. 10. New York: Elsevier; 2011

[8] Bousquet et al. Introduction to Statistical Learning. 2003. Available from: http://www.econ.upf.edu/~lugosi/mlss_slt.pdf

[9] Field A. Discovering Statistics Using R. London: Sage; 2012

[10] Carmichael I, Marron JS. Data science vs. statistics: Two cultures? Japanese Journal of Statistics and Data Science. 2018;**1**(1):117-138

[11] Cahn A, Shoshan A, Sagiv T, Yesharim R, Goshen R, Shalev V, et al. Prediction of progression from pre-diabetes to diabetes: Development and validation of a machine learning model. Diabetes/Metabolism Research and Reviews. 2020;**36**(2):e3252. DOI: 10.1002/dmrr.3252 Epub 2020 Jan 14

[12] Breiman L. Statistical modeling: The two cultures (with comments and a rejoinder by the author). Statistical Science. 2001;**16**(3):199-231

[13] Mehta R, Uppunuthula S. Use of machine learning techniques to identify the likelihood of hospitalization for high-risk patients diagnosed with COVID-19. In: ISPOR Conference; Washington DC. 2022

[14] American Diabetes Association. Understanding A1C Diagnosis. 2022. Available from: https://www.diabetes.org/diabetes/a1c/diagnosis#:~:text=Diabetes%20is%20diagnosed%20at%20fasting,equal%20to%20126%20mg%2Fdl

[15] Bzdok et al. Machine learning: A primer. Nature Methods. 2017;**14**(12):1119-1120. DOI: 10.1038/nmeth.4526

[16] Bellman RE. Adaptive Control Processes. Princeton, NJ: Princeton University Press; 1961

[17] Hastie T, Tibshirani R, Friedman J. The Elements of Statistical Learning: Data Mining, Inference, and Prediction (2ed). Stanford, CA: Springer; 2016

[18] Chapman et al. Statistical learning theory for high dimensional prediction: Application to criterion-keyed scale development. Psychology Methods. 2016;**21**(4):603-620. DOI: 10.1037/met0000088

[19] Argent et al. The importance of real-world validation of machine learning systems in wearable exercise biofeedback platforms: A case study. Sensors (Basel). 2021;**21**(7):2346. DOI: 10.3390/s21072346

[20] Parikh et al. Understanding and using sensitivity, specificity and predictive values. Indian Journal of Ophthalmology. 2008;**56**(1):45-50. DOI: 10.4103/0301-4738.37595

[21] Mendis A. Statistical Modeling vs. Machine Learning. 2019. Available from: https://www.kdnuggets.com/2019/08/statistical-modelling-vs-machine-learning.html

[22] Hayes K, Rajabathar R, Balasubramaniam V. Uncovering the machine learning "Black Box": Discoveringlatent patient insights using text mining & machine learning. In: Conference Paper Presented at Innovation in Analytics via Machine Learning & AI; Las Vegas, NV. 2019 Available from: https://www.pmsa.org/other-events/past-symposia

[23] Belabbas M, Wolfe PJ. Spectral methods in machine learning and new strategies for very large datasets. Proceedings of the National Academy of Sciences. 2009;**106**(2):369-374. DOI: 10.1073/pnas.0810600105

[24] Kempa-Liehr et al. Healthcare pathway discovery and probabilistic machine learning. International Journal of Medical Informatics. 2020;**137**:104087. DOI: 10.1016/j.ijmedinf.2020.104087

[25] Wasserman L. Rise of the machines. In: Past, Present, and Future of Statistical Science. Chapman and Hall; 2013. pp. 1-12. DOI: 10.1201/b16720-49

[26] Ranjan R. Calibration in machine learning. 2019. Available from: https://medium.com/analytics-vidhya/calibration-in-machine-learning-e7972ac93555

[27] Child CM, Washburn NR. Embedding domain knowledge for machine learning of complex material systems. MRS Communications. 2019;**9**(3):806-820. DOI: 10.1557/mrc.2019.90

[28] Hilliermeir E, Waegerman W. Aleatoric and epistemic uncertainty in machine learning: An introduction to concepts and methods. Machine Learning. 2021;**110**:457-506. DOI: 10.1007/s10994-021-05946-3

[29] Goh et al. Evaluating human versus machine learning performance in classifying research abstracts. Scientometrics. 2020;**125**:1197-1212. DOI: 10.1007/s11192-020-03614-2

[30] Chicco D, Jutman G. The advantages of the Matthews correlation coefficient (MCC) over F1 score and accuracy in binary classification evaluation. BMC Genomics. 2020;**21**(6). DOI: /10.1186/s12864-019-6413-7

[31] Ansari et al. Rethinking human-machine learning in Industry 4.0: How does the paradigm shift treat the role of human learning? Procedia Manufacturing. 2018;**23**:117-122. DOI: 10.1016/j.promfg.2018.04.003

[32] Morganstein et al. Predicting population health with machine learning: A scoping review. BMJ Open. 2020;**10**(10). DOI: 10.1136/bmjopen-2020-037860

[33] Terranova et al. Application of machine learning in translational medicine: Current status and future opportunities. The AAPS Journal. 2021;**23**(74). DOI: 10.1208/s12248-021-00593-x

[34] Kleczyk E, Hayes K, Bennett M. Building organization AI and ML acumen during the COVID Era. 2022. In: PMSA Annual Conference. Louisville, KY. 2022

Chapter 3

Image-Based Crop Leaf Disease Identification Using Convolution Encoder Networks

Indira Bharathi and Veeramani Sonai

Abstract

Nowadays, agriculture plays a major role in the progress of our nation's economy. However, the advent of various crop-related infections has a negative impact on agriculture productivity. Crop leaf disease identification plays a critical role in addressing this issue and educating farmers on how to prevent the spread of diseases in crops. Researchers have already used methodologies such as decision trees, random forests, deep neural networks, and support vector machines. In this chapter, we proposed a hybrid method using a combination of convolutional neural networks and an autoencoder for detecting crop leaf diseases. With the help of convolutional encoder networks, this chapter presents a unique methodology for detecting crop leaf infections. Using PlantVillage dataset, the model is trained to recognize crop infections based on leaf images and achieves an accuracy of 99.82%. When compared with existing work, this chapter achieves better results with a suitable selection of hyper tuning parameters of convolution neural networks.

Keywords: crop leaf, convolution neural network, autoencoder, ReLU, deep neural network, hyper tuning

1. Introduction

In agriculture, crop leaf plays an important role in giving information about the good growth of the plant. Various climatic factors affect the growth of the plant. Besides natural calamities, crop leaf disease is a major hazard to the growth of agriculture yields and economic victims. Once we fail to analyze the infections in the crops, we may lead to low pesticide usage. Therefore, crop leaf identification is considered a major issue in the biological features of diseases present in the crop. When required, expert visual inspections and biological reviews are normally carried out through plant diagnosis. This strategy, on the other hand, is usually time-consuming and ineffective. To solve these difficulties, sophisticated and intelligent systems for detecting plant diseases are required.

To provide an intelligent system to identify the crop leaf diseases, we proposed a convolution neural network with image processing methodologies such as image segmentation and filtering. In the existing works, most researchers applied conventional machine learning algorithms to predict or identify the crop diseases present in the leaf. However, machine learning algorithms better recognize the plants, weed discrimination, etc. As a result, crop leaf disease identification is critical to maintaining agricultural productivity. In general, plant leaf disease analysis is also done manually by using visual inspection. But it is time-consuming and potentially error-prone. As a result, diagnosing crop disease using automated procedures is beneficial since it reduces a significant amount of effort associated with crop monitoring on large farms, and it detects disease symptoms at an early stage, i.e., when the disease first appears. Leaf plant health monitoring and early detection of symptoms are required to limit disease transmission, which aids farmers in effective management methods.

To develop an accurate image classifier for crop leaf identification, we need image samples of damaged and healthy crops. The PlantVillage dataset has thousands of images of healthy and infected crop leaves. In this dataset, six diseases in three crop species are labeled. Hence, we use 54,306 image samples with a convolution encoder network to identify the crop leaf diseases more accurately. The main contribution of this chapter is summarized as follows:

1. A brief review of convolution neural network has been conducted to identify diseases in several crops/plants affected by fungi, viruses, etc.

2. Features are extracted by using an encoder, namely variational autoencoder.

3. An effort has been made to improve the performance of CNN for identifying the crop leaf by using segmentation of the images.

The rest of the chapter is organized as follows: The literature is briefly explained in Section 2. The techniques used have been elaborated in Section 3. The results were elaborated in Section 4, and the conclusion and future work are provided in Section 5.

2. Related works in the literature

An existing literature survey categorized the plant diseases by using several Convolutional Neural Networks (CNN) [1, 2]. In the PlantVillage dataset, another CNN-based architecture was presented to classify disease, and it outperformed DL models such as as AlexNet, VGG-16, Inception-v3, and ResNet [3]. CNN model is also proposed in a study to classify data in tea leaves. A CNN-based approach was developed for groundnut disease categorization in a recent publication [4]. Similarly, little literature has looked at sophisticated training strategies; for example, [5] focused at the performance of AlexNet and GoogLeNet. By comparing state-of-the-art and fine-tuning techniques, comparison research was undertaken to demonstrate the importance of the fine-tuning technique.

A random forest-based classifier to identify the healthy and affected leaf is proposed [6]. The author has described the dataset creation, extraction, and training. An AlexNet classification technique is applied to detect rice leaf diseases, namely bacterial blight, brown spot as well as leaf smut [7]. In order to monitor regularly and automatic disease detection for remote sensing images was proposed [8]. Using Canny's edge detection and machine learning algorithm, a disease identification system was proposed [9]. A convolutional neural networks-based autoencoder was used to detect crop leaf diseases. The convolution filter size of 2×2 and 3×3 gives different accuracy for the different eoches [10].

A state-of-the-art deep convolutional neural network for image classification is proposed in [11]. A DenseNet model is proposed to perform better than other models. The author proposes activation functions that perform better than ReLU on various scales [12]. For the early detection of European wheat diseases, an automatic plant disease diagnosis system is proposed [13, 14]. To increase the robustness of crop detection, a multi-target tracking algorithm is proposed [15].

In order to classify the leaf images, deep learning approaches are studied [16]. For the leaf segmentation, the images are trained using Mask Regionased Convolutional Neural Network (Mask R-CNN). The average accuracy obtained for the VGG16 images is 91.5%. Through deep learning methodologies, leaf images are classified as healthy and affected [17]. A method to dynamically analyze the images of the disease is proposed in [18]. The output is sent to the farmer, and the feedback is reflected in the model. Using the deep learning, strawberry fruits and leaves, diseases are diagnosed [18]. A convolutional Neural Network (CNN) model and Learning Vector Quantization (LVQ) algorithm-based method for tomato leaf disease detection and classification [19, 20].

To categorize the healthy and affected leaf, a deep learning model is applied over the public images [1]. For the sustainable development of arming, it is essential to use Artificial intelligence and machine learning approaches [21]. To solve the current agricultural problems, a computer vision technology is combined with deep learning model [22]. Using the images of plants, a state-of-the-art deep learning model is applied to detect disease [5, 23]. To enhance the accuracy, a depthwise separable convolution is adopted [3]. For the automatic detection of infection in the tomato leaves, an enhanced deep learning architecture is adopted over the plantVillage datase [4]. To classify the crop, a novel three-dimensional (3D) convolutional neural network (CNN) is applied over the remote sensing image [24].

3. Proposed hybridized convolution neural network with variational autoencoders system

In this chapter, a hybridized convolution neural network with variational autoencoders is proposed to classify the crop leaf diseases, and hence, it is named as V-Convolution encoder network. To extract the informative features of the leaf, we used an autoencoder. It is a type of neural network, which is useful for outperforming two functions, namely encoding and decoding. An encoding part plays a role in extracting the high-dimensional features of the leaf, and the decoding part reconstructs the inputs taken. In general, all CNN consists of three important layers, namely encoder layers, max or min – pooling layers, and fully connected layers, as shown in **Figure 1**.

3.1 Building blocks of CNN

The convolutional layer is the core part of a convolutional network that contains a structure of learnable channels. In the forward pass, the width and height information of the images is passed over each channel, and the product of kernel and image pixels is calculated. In the backward pass, the gradient of the loss with corresponds to input, weight, and bias is computed. The various levels of filters are used to extract the needed features from the matrix of original images taken. As the filter levels go in deep, we can solve a more specific problem. To hold the important features, zero padding is added across the image matrix.

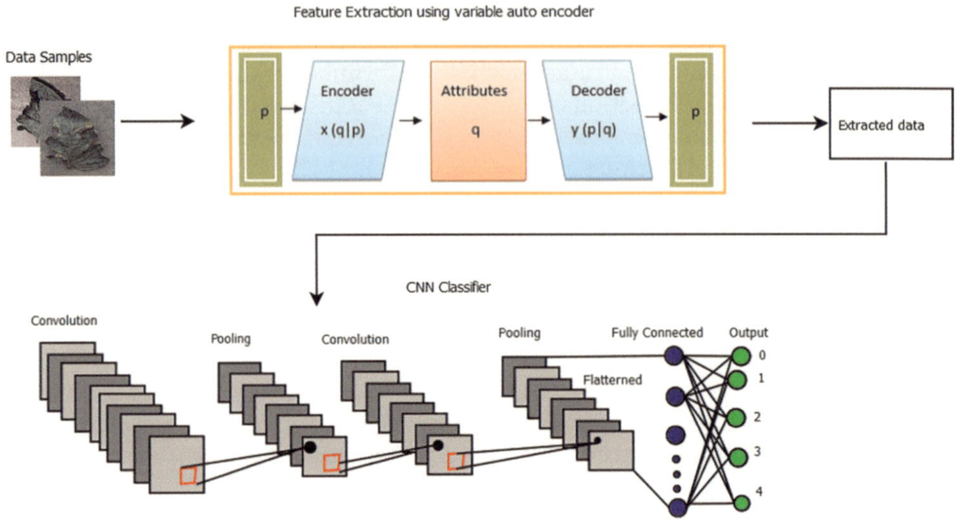

Figure 1.
Proposed architeure.

The ReLU activation function is used within the convolution layer, which adds nonlinearity to the network. It calculates the weighted informative features faster than the tangent or sigmoidal function. Next is the max pooling layer, which increases a pooling layer in the midstream of several convolutional layers. Its skill is to vigorously decrease the spatial size of the image to minimize the size of parameters and calculation and consequently to control overfitting. When the image size is large, the pooling layer reduces the number of training features. The important significance of adding pooling layers is to lessen the spatial size of the input image. Here, min-max pooling is used in our implementation. After the pooling layer, the fully connected layer is essential to produce an output equivalent to the number of classes that we want as output. In this, the annihilation of neurons is done, and we gain a vector of all neurons. In such a layer, all neurons are fully connected with neurons in the previous layer. At last softmax layer is used to calculate the probabilities should be in the range 0–1, and the summation of all probabilities is 1.

3.2 Variational autoencoder

Variational autoencoder is proposed to extract the features of given input images. It is a neural organization that is intended for unsupervised learning. It comprises two sections: encoder and decoder. The encoder means to encode input highlights into encoding vectors, while the decoder acquires the yield highlights back from the encoding vector. The encoder is planned so that the result produces a variable, which is a compressed form of the input. On the other side, the decoder decompresses the resultant images back to their original size. The difference between autoencoder and variation autoencoder is that the autoencoder represents the features by applying the function, whereas the variational autoencoder represents the features by calculating the probability distribution. This encoder is designed based on the principle of a neural network that gives q of input as p of output. In a probability distribution model, this network parameterizes the inaccurate features of the input images q and produces the result as a distribution of x (p | q). This variational decoder then reconstructs the input samples p such that it produces parameters to the distribution y (p | q). This model consists of two phases, namely feature learning and classifier. The learning of features is done in an unsupervised network, whereas the leaf diseases are classified by training the samples using a CNN classifier. The overall architecture of the proposed system is shown in **Figure 1**.

To perform the crop leaf disease identification, we have considered the PlantVillage dataset. To improve the performance of the proposed system, the segmentation process is performed on the original data samples before feature learning (**Table 1**).

Class	Diseases	Crop
0	Bacterial_spot	Pepper, Tomato
1	Target_Spot	Tomato
2	Early_blight	Potato, Tomato
3	Late_Blight	Potato, Tomato
4	Leaf_Mold	Tomato
5	mosaic_virus	Tomato
6	Healthy	All (Pepper, Potato and Tomato)

Table 1.
Classes of various crop diseases.

PlantVillage Dataset					
Classes of Tomato leafs	**Label_Name**	**Disease type**	**Sample images**		
			Training	**Validation**	**Testing**
Tomato_Target_Spot	Tom_target	Fungi	994	270	140
Tomato_mosaic_virus	Tom_mosaic	virus	266	70	37
Tomato_YellowLeaf_Curly_Virus	Tom_curly	virus	3786	1071	500
Tomato_Bacterial_Spot	Tom_bact	Bacteria	1494	420	219
Tomato_Early_Blight	Tom_EBlight	Fungi	250	150	100
Tomato_Healthy	Tom_Hlthy	—	1128	310	153
Tomato_Late_Blight	Tom_LBlight	Bacteria	700	200	100
Tomato_Leaf_Mold	Tom-mold	Fungi	667	195	90
Tomato_Septoria_leaf_spot	Tom_septo	Fungi	1247	354	170
Tomato_Spider_mites	Tom_spider	Mite	1181	330	165

Table 2.
Training, test, and validation values used for each category of data sets.

We acquired our results based on the training and testing sample listed in **Table 2**. For classification, we considered different types of crop diseases from tomato leaves. **Table 1** describes the various crop and their diseases. It has six different classes, ranging from 1 to 6. The proposed network has been trained to recognize crop infections based on leaf images. Different convolution filter levels are used in the proposed work, and to train the network more efficiently, ReLU activation function is used. It was shown that the proposed architecture achieved better accuracy for various epochs and convolution filter sizes. We applied additional convolution layers with 128 filters and filtered size 2 × 2 with ReLU. It is then followed by two additional convolution layers with 256 filters and filter size 2 × 2 with ReLU. After all this, a flattening layer is used to acquire a vector of neurons that uses ReLU function. Then two dense layers are used: one uses ReLU, while the other uses the softmax function and depicts the output class.

4. Results and discussion

The proposed system is implemented by using Python Scikit-learn packages and is executed using the Intel i5 processor. The proposed approach is evaluated by using the PlantVillage dataset [25]. The testing and training used for the leaf image dataset are illustrated in **Table 2**. The following performance parameters have been considered for our implementation, namely precision, recall, and F1-score. The results are taken with different values of epochs, and it is compared with existing approaches. By varying the epochs, the error in the testing and training sample is plotted in **Figure 2**.

We achieved 98% of accuracy if the network is iterated for 150 epochs. It is also observed that as the filter size increases, we get 100% accuracy. **Table 2** shows the training and testing accuracy for the different convolutional filter sizes such as 2 × 2 and 3 × 3. The best training accuracy for the 2 × 2 filter size is 97.21, and the best testing accuracy is 87.12 for filter size 2 × 2. When compared with existing work, this paper achieves better results with a suitable selection of hyper tuning parameters of a convolution neural network (**Table 3**).

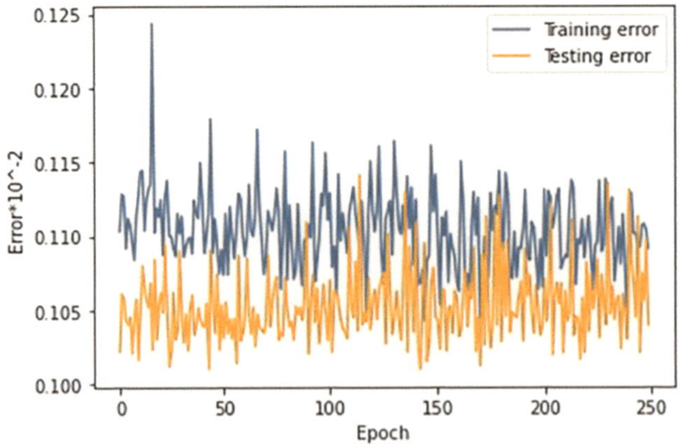

Figure 2.
Comparison of training and testing error with various epoches.

Epochs	Convolution filter	Training accuracy	Testing accuracy
100	2 *2	95.3	82.71
100	3*3	98.34	85.45
150	2*2	92.21	87.12
150	3*3	97.73	84.78
200	2*2	94.67	80.71
200	3*3	95.18	81.1

Table 3.
Training and testing accuracy for different filters.

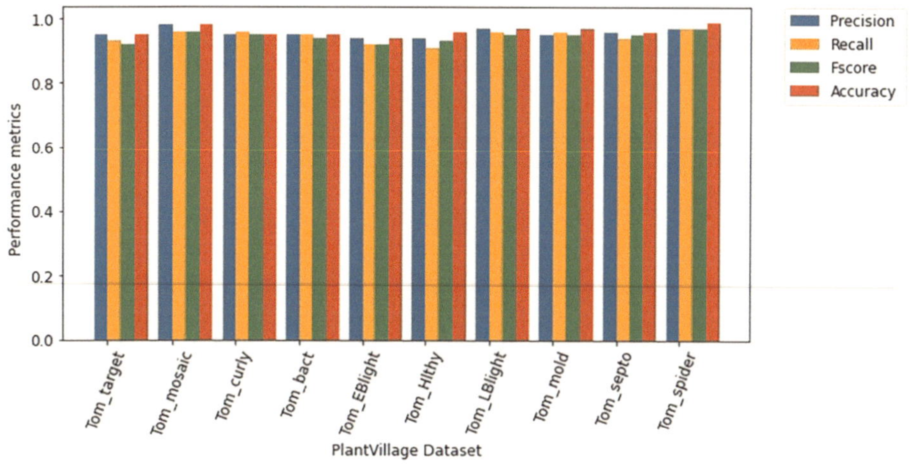

Figure 3.
Comparison of various performance parameters.

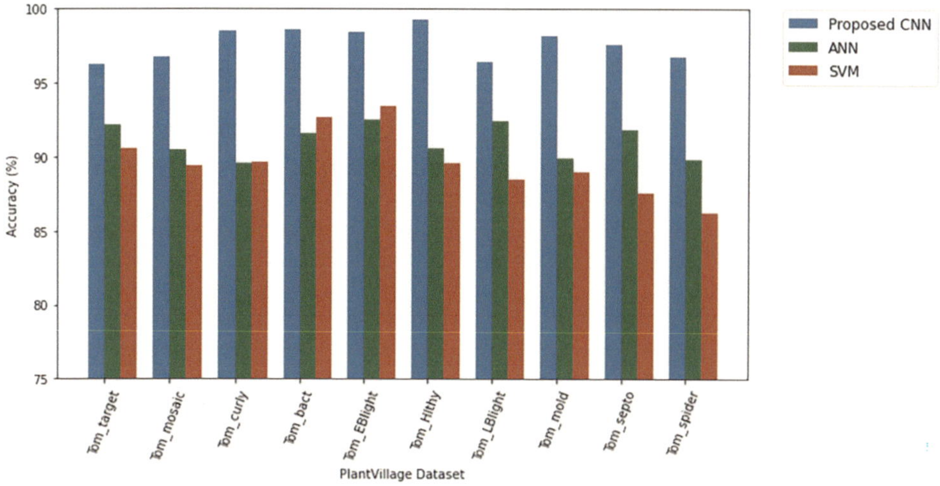

Figure 4.
Accuracy comparison of various classifier.

No. of Epochs	Performance metrics	Proposed method									
		Tom_target	Tom_mosaic	Tom_curly	Tom_bact	Tom_EBlight	Tom_Hlthy	Tom_LBlight	Tom-mold	Tom_septo	Tom_spider
100	Precision	0.94	0.93	0.94	0.93	0.90	0.89	0.96	0.93	0.93	0.85
	Recall	0.92	0.94	0.93	0.92	0.91	0.91	0.94	0.94	0.92	0.84
	F1-Score	0.92	0.92	0.93	0.93	0.91	0.91	0.94	0.94	0.92	0.87
	Accuracy	0.95	0.95	0.93	0.93	0.91	0.92	0.95	0.93	0.91	0.89
150	Precision	0.94	0.97	0.96	0.96	0.93	0.91	0.98	0.96	0.95	0.99
	Recall	0.93	0.96	0.96	0.95	0.92	0.91	0.96	0.96	0.94	0.97
	F1-Score	0.92	0.96	0.95	0.94	0.92	0.93	0.95	0.95	0.95	0.97
	Accuracy	0.95	0.98	0.95	0.95	0.94	0.94	0.97	0.95	0.96	0.97
200	Precision	0.93	0.95	0.96	0.94	0.92	0.90	0.96	0.94	0.94	0.99
	Recall	0.93	0.95	0.95	0.93	0.92	0.89	0.94	0.94	0.93	0.97
	F1-Score	0.92	0.96	0.95	0.93	0.92	0.89	0.96	0.96	0.93	0.97
	Accuracy	0.92	0.96	0.95	0.95	0.91	0.89	0.95	0.94	0.93	0.97

Table 4.
Precision, Recall, F1-score, and Accuracy value of various datasets.

The performance of the resulting implementation is illustrated in the **Figure 3**. **Figure 4** shows the comparison of the proposed classifier and existing classifier approaches. The proposed CNN approach shows superior performance in terms of accuracy compared with other existing approaches (**Table 4**).

5. Conclusion and future work

Crop leaf diseases have been responsible for reducing production resulting in economic causes. Recently, the crop leaf has been facing several diseases from various insects and pests. This chapter proposes a unique methodology for detecting crop leaf infections. With the PlantVillage dataset, the model is trained to recognize crop infections based on leaf images and achieves an accuracy of 99.82%. This chapter presented a feature selection algorithm to identify essential features from crop leaf images. The chosen features are given to the hybrid method using a combination of convolutional neural networks and autoencoders. Among all the existing classifiers, the proposed approach shows an average of 84.54% of execution time improvement in performing the classification. This work can be enhanced further to give the recommendation to the farmer to apply proper insecticides prior to the spread of such diseases.

Author details

Indira Bharathi* and Veeramani Sonai
School of Computing, Amrita Vishwa Vidhyapeetham, Vengal, Tamil Nadu, India

*Address all correspondence to: b_indira@ch.amrita.edu

References

[1] Kulkarni O. Crop disease detection using deep learning. In: IEEE Fourth International Conference on Computing Communication Control and Automation (ICCUBEA). New York: IEEE; 2018. pp. 1-4

[2] Kaushik M, Prakash P, Ajay R, Veni S. Tomato leaf disease detection using convolutional neural network with data augmentation. In: IEEE 2020 5th International Conference on Communication and Electronics Systems (ICCES). New York: IEEE; 2020. pp. 1125-1132

[3] Kamal K, Yin Z, et al. Depthwise separable convolution architectures for plant disease classification. Computers and Electronics in Agriculture. 2019;**165**: 104948

[4] Karthik R, Hariharan M, Anand S, Mathikshara P, Johnson A. Attention embedded residual CNN for disease detection in tomato leaves. Applied Soft Computing. 2020;**86**:105933

[5] Bedi P, Gole P, Agarwal SK. Using deep learning for image-based plant disease detection. In: Internet of Things and Machine Learning in Agriculture. Lausanne, Switzerland: Frontiers; 2021. pp. 369-402

[6] Maniyath SR, Ram H. Plant disease detection using machine learning. In: IEEE International Conference on Design Innovations for 3Cs Compute Communicate Control (ICDI3C). New York: IEEE; 2018. pp. 41-45

[7] Matin MH, Uddin MS. An efficient disease detection technique of Rice leaf using AlexNet. Journal of Computer and Communications. 2020;**8**(12):4

[8] Badage A. Crop disease detection using machine learning: Indian Agriculture. International Research Journal of Engineering and Technology. 2018;**5**(9):866-869

[9] Kranth PR, Lalitha H, Basava L, Mathur A. Plant disease prediction using machine learning algorithm. International Journal of Computer Applications. 2018;**182**(25):41-45

[10] Khamparia A, Saini G, Gupta D, Khanna A, Tiwari S. Seasonal crops disease prediction and classification using deep convolutional encoder network. Circuits, Systems and Signal Processing. 2020;**39**:818-836

[11] Too EC, Yujian L, Njuki S, Yingchun L. A comparative study of fine-tuning deep learning models for plant disease identification. Computers and Electronics in Agriculture. 2018;**22**:135-152

[12] Qian S, Liu H, Liu C, Wu S, Wong HS. Adaptive activation functions in convolutional neural networks. Neurocomputing. 2018;**272**:204-212

[13] Picon A, Alvarez-Gila A, Seitz M, Ortiz-Barredo A, Echazarra J, Johannes A. Deep convolutional neural networks for mobile capture device-based crop disease classification in the wild. Computational Electronic Agriculture. 2018;**138**:200-209

[14] Lu J, Hu J, Zhao G, Mei F, Zhang C. An in-field automatic wheat disease diagnosis system. Computers and Electronics in Agriculture. 2017;**142**: 369-379

[15] Hamuda E, Mc Ginley B, Glavin M, Jones E. Improved image processing-based crop detection using Kalman filtering and the Hungarian algorithm. Computers and Electronics in Agriculture. 2018;**148**:37-44

[16] Yang K, Zhong W. Leaf segmentation and classification with a complicated background using deep learning. Agronomy. 2020;**10**:1721

[17] Ferentinos KP. Deep learning models for plant disease detection and diagnosis. Computers and Electronics in Agriculture. 2018;**145**:311-318

[18] Park H, JeeSook E, Kim S-H. Crops disease diagnosing using image-based deep learning mechanism. In: International Conference on Computing and Network Communications (CoCoNet). New York: IEEE; 2018. pp. 23-26

[19] Sardogan M, Tuncer A, Ozen Y. Plant leaf disease detection and classification based on CNN with LVQ algorithm. In: IEEE International Conference on Computer Science and Engineering (UBMK). New York: IEEE; 2018. pp. 382-385

[20] Reddy JN, Vinod K, Ajai AR. Analysis of classification algorithms for plant leaf disease detection. In: IEEE International Conference on Electrical, Computer and Communication Technologies (ICECCT). New York: IEEE; 2019. pp. 1-6

[21] Jha K, Doshi A, Patel P, Shah M. A comprehensive review on automation in agriculture using artificial intelligence. Artificial Intelligence Agriculture. 2019; **2**:1-12

[22] Tian H, Wang T, Liu Y, Qiao X, Li Y. Computer vision technology in agricultural automation – A review. Agriculture. 2020;**7**(1):1-19

[23] Too EC, Yujian L, Njuki S, Yingchun L. A comparative study of fine-tuning deep learning models for plant disease identification. Computers and Electronics in Agriculture. 2019;**161**:272-279

[24] Ji S, Zhang C, Xu A, Shi Y, Duan Y. 3D convolutional neural networks for crop classification with multi-temporal remote sensing images. Remote Sensing. 2018;**10**(1):75-84

[25] https://knowyourdata-tfds.with google.com/

Chapter 4

Perspective Chapter: Deep Reinforcement Learning for Co-Resident Attack Mitigation in The Cloud

Suxia Cui and Soamar Homsi

Abstract

Cloud computing brings convenience and cost efficiency to users, but multiplexing virtual machines (VMs) on a single physical machine (PM) results in various cybersecurity risks. For example, a co-resident attack could occur when malicious VMs use shared resources on the hosting PM to control or gain unauthorized access to other benign VMs. Most task schedulers do not contribute to both resource management and risk control. This article studies how to minimize the co-resident risks while optimizing the VM completion time through designing efficient VM allocation policies. A zero-trust threat model is defined with a set of co-resident risk mitigation parameters to support this argument and assume that all VMs are malicious. In order to reduce the chances of co-residency, deep reinforcement learning (DRL) is adopted to decide the VM allocation strategy. An effective cost function is developed to guide the reinforcement learning (RL) policy training. Compared with other traditional scheduling paradigms, the proposed system achieves plausible mitigation of co-resident attacks with a relatively small VM slowdown ratio.

Keywords: cloud computing, risk mitigation, resource management, co-resident attack, reinforcement learning

1. Introduction

Cloud Computing, which has its origins in expanding the Internet, aims to provide remote and scalable computing and storage resources to its customers. Users from small businesses in a locally resource-limited environment can manipulate and store large datasets for real-time processing with cloud services. The cloud platform has gradually reshaped daily lives because it has been recognized as a convenient way to transmit and store data in the big data era. Organizations can choose from the public, private, or hybrid cloud that combines the public and private deployment model features. The term "XaaS" is coined for the service-oriented architecture, emphasizing that anything can be treated as a service under the cloud computing environment. Examples of cloud delivery services include infrastructure as a service (IaaS),

Approved for Public Release on 01 June 2022; Distribution Unlimited; Case number: AFRL-2022-2581.

platform as a service (PaaS), and software as a service (SaaS) [1]. Recently, function as a service (FaaS) further expanded the backend as a service (BaaS) offering. Under each delivery model, a cloud service provider (CSP) is responsible for allocating enough resources to maintain quality of service (QoS) to the users and protect their data from security risks.

Virtualization has been adopted by most cloud computing platforms to profit from the "pay as you go (PAYG)" model [2]. Virtualization is an idea generated from IBM's mainframe platform in the early 1960s. After entering the twenty-first century, it was successfully utilized in cloud computing that can bring down the cost of maintaining a large-scale system. It converts a physical server into numerous VMs, rented out to several occupants [3–7]. This VM-PM relationship is illustrated in **Figure 1**.

The apparent relationship between PMs running and power consumption places a high demand on a strategy for energy minimization in this configuration [8]. Security and data privacy are other concerns for cloud computing platforms. Attackers will seek to exploit any vulnerability to achieve various malicious goals on the victim's network, software, and databases. The co-resident attack is one of the prevalent cybersecurity risks resulting from virtualization. Ideally, two neighboring VMs are isolated from each other when running their tasks. However, in reality, each co-located VM will depend on the same PM where hardware, like CPUs or memory elements, is shared by all the VMs. Therefore, a VM's private information may be accessed by its neighboring VM by launching side-channel attacks [9–12], as shown in **Figure 2**. Here, a hypervisor or virtual machine monitor (VMM) creates and runs VMs on a hosting PM. The arrows illustrate the route of side-channel attacks. A side-channel attack is a significant security challenge that prevents many organizations from adopting cloud computing technology. Although recently deep learning algorithms have proven to be effective in cloud resource management [13–16], few paid attention to side-channel attack avoidance at the same time.

To fill in this gap, we developed a novel deep reinforcement-learning (DRL) based dynamic VM allocation approach to optimize the trade-offs between the VM completion and the co-resident risks mitigation. The main contributions of this paper are as follows:

- Threat model design: A time-sensitive zero trust threat model is developed for co-resident vulnerability analysis. The model enables the tracking of VM co-existent pairs on the same PM.

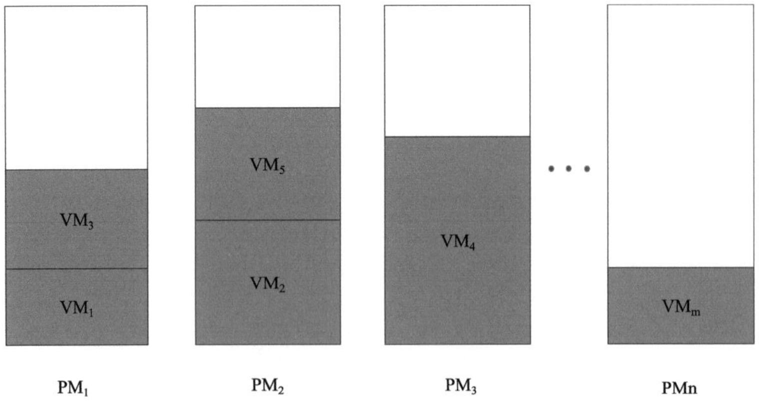

Figure 1.
VMs and PMs in a data center via virtualization.

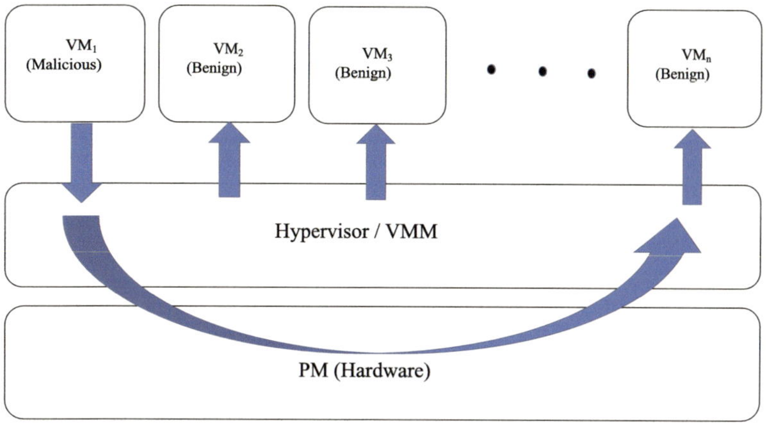

Figure 2.
Side-channel attacks in cloud computing.

- DRL adoption to co-resident risk mitigation: This article is the first to investigate the different states, actions, and rewards of DRL to fit it into cloud computing side-channel attacks. The rewards guide the VMM decision-making.

- Simulation platform implementation: The proposed system accepts tasks dynamically at runtime and analyzes cloud co-resident risks at each timestep for optimized scheduling algorithms.

2. Related works: Secure cloud resource management

Resource management alone without security consideration in a cloud computing environment is already very challenging. Multi-tenant environments allow attackers to begin a co-resident infiltration and steal the victim's information by side channels. The risks that attackers pose to VMs on the same hypervisor are growing security concerns and are being addressed differently. This section will discuss established current resource management approaches with and without security awareness. Three categories of methodologies are commonly used: Heuristic, game theory, and machine learning (ML).

The exploration of heuristics and meta-heuristics to solve nondeterministic polynomial time (NP) problems are growing amid the difficulties to solve them using traditional methods [17]. Gawali and Shinde [18] combined a modified analytic hierarchy process (MAHP), bandwidth aware divisible scheduling (BATS), and the longest expected processing time preemption (LEPT) to achieve improved performance. Qin *et al.* [20] took the idea of "ant colony optimization" from [19] and proposed a probabilistic algorithm that can simultaneously maximize the revenue of communications and minimize the power consumption of PMs. Similarly, Tawfeek *et al.* also adopted ant species' nature and presented a random optimization search approach for allocating the incoming jobs to the virtual machines [21]. The proposed method outperformed the popular first come first serve method. Patel introduced a hybrid algorithm that used a modified honeybee behavior-inspired algorithm for priority-based tasks and an enhanced weighted round-robin algorithm for non-priority-based tasks [22] to balance the workload over the cloud dynamically. When using heuristic methods with the consideration of co-resident attacks, various policies were compared. Jia *et al.* [23] proposed a VM allocation method to optimize load balancing and reduce energy consumption and security risks by managing CPU utilization of the hosting PMs. Miao *et al.* offered two metrics to outline co-residency and conflict of the cloud [24]. Both placement and migration algorithms mediate differences between tenants to alleviate co-resident attacks in the cloud proactively. Han *et al.* formulated a set of security metrics and a quantitative model to assign new VMs to the server with the most VMs [25]. The research uncovered that the server's configuration, oversubscription, and background traffic had a substantial impact on the ability to stop attackers from co-locating with the targets.

Yang *et al.* [26] explored a simplified algorithm for energy management in cloud computing. The paper centered around establishing a mathematical model to calculate computing nodes' stability, configuring a game-theoretic cooperative model for the task of scheduling cloud computing, and examining the problem as a multi-stage sequential game. Patra *et al.* presented the task as a player and the VM as a strategy in [27]. A non-cooperative game scheduling and a task balance scheduling algorithm are compared to collect the node's average task processing speed. Therefore, it was determined that the game-theoretic algorithm proposed could improve energy management in cloud computing. In [28], the cooperative behavior of multiple cloud servers was studied. An evolutionary mechanism was presented in the hierarchical cooperative game model for VMs deployment strategy to improve the efficiency in the public cloud environment. Jia *et al.* modeled several basic VM allocation policies using game theory to achieve a quantitative analysis, while also presenting the attack effectiveness, coverage, power consumption, workload balance, and cost under the VM allocation policies and solving the mathematical solution in CloudSim [23]. Their results found that to reduce the efficiency rates for the attacker, the cloud provider should apply a probabilistic VM allocation policy. Narwal *et al.* proposed a payoff matrix and a decision tree for any number of users [29, 30]. When a unique user was selected, the choices of investing in security were assessed until equilibrium was reached. Security games are a way of blocking the attacker's ability to locate the VMs they are searching for. Han *et al.* proposed a policy pool with multiple VM allocation policies from which to select the policy that will be used with a certain probability [31].

Difficulties regarding energy efficiency in cloud computing can also be addressed using machine learning-based techniques [32]. Witanto *et al.* employed a neural network-based adaptive selector procedure to arrange the VMs on the physical servers in data centers [33]. Pahlevan *et al.* presented a hyper-heuristic algorithm to exploit both heuristic and ML-based VM allocation methods by selecting the best one during run-time [32]. Zhang *et al.* [34] suggested an auction-based resource allocation scheme to represent a machine learning classification or regression problem. They outlined machine learning classification and posed two resource allocation prediction algorithms rooted in linear and logistic regression. Liu *et al.* presented a reinforcement learning-based approach to allow complex scenarios to efficiently manage resources [35]. In order to do so, they used neural networks to grasp the goal of the research model, RL to enhance the model, and E-greedy methodology to expand the RL process. Their approach lowered job delay for hybrid scenarios. ML-based methods have been proposed to fight against co-resident attacks focusing on different factors, such as minimizing the time of a malicious VM co-location. Joseph *et al.* [36] used traditional ML algorithms, such as support vector machine (SVM), naïve bayes, and random forests to detect malware, following a self-healing methodology to power off the attacked VMs and restore them to healthy conditions. In reality, there is a concern with the amount of time it takes to implement a solution to mitigate VMs in the event of co-resident attacks. To the best of our knowledge, no current ML-based approach succeeded in mitigating co-resident attacks based on VM mitigation, while minimizing the VM downtime.

3. Threat model

There are many approaches to fight against co-resident attacks, including hardware modification, intrusion detection, secure VM allocation, and migration. Threat model building is crucial to guide proper defense. This section goes through the study of existing models and presents our proposed threat model with detailed variable selections.

3.1 Modeling co-resident attacks

Many optimization models were proposed to fight against co-resident attacks. Abazari *et al.* suggested a multi-objective optimization method to calculate alternative responses with the least amount of threat through graphics and proper attack countermeasures [37]. Liu *et al.* considered the three main factors which lead to the likelihood of malicious VMs co-locating with normal users [38]. Berrima *et al.* used a VM placement strategy to reduce the co-location attacks with complete resource optimization. Their approach presents a trade-off between security and VM startup delay [39]. Hasan *et al.* proposed a co-resident attacks mitigation and prevention (CAMP) model to separate malicious and benign VMs by comparing existing models over data security, data survivability, and user storage overhead [5]. Other works focused on a probabilistic co-residence coverage optimization model, while combining a data partition technique that involves arranging servers randomly [40, 41].

3.2 Proposed threat model with detailed design components

Our proposed approach takes the time-sensitive risk level from co-resident attacks into account and searches for the solution to the dynamic VM allocation problem through DRL. Research shows that the co-resident attack will have a total cycle of t_3, consisting of three stages: probe, construct, and launch. Probe and construct generate a configuration interval. This is illustrated in **Figure 3**. To avoid the attacks, the defender must take action before the launching starts. In other words, before the configuration interval t_2 is reached [42].

Our co-resident risk model is developed in a similar scenario to [43]. The choice of variables is listed in **Table 1**.

The co-resident risk indicator can be obtained through the following equations:

$$rcr(v_i, v_j) = t_s(v_i) \times CoRes(v_i, v_j) \times t_s(v_j) \tag{1}$$

$$CoResFactor = \begin{cases} \alpha_0 & \text{for } CoRes(v_i, v_j) < t_1 \\ \alpha_1 & \text{for } CoRes(v_i, v_j) \in [t_1, t_2) \\ \alpha_2 & \text{for } CoRes(v_i, v_j) \geq t_2 \end{cases} \tag{2}$$

Variables	Descriptions
N	No. of PMs
n	No. of resource slots in one PM
M	No. of VMs
X	The mapping between VMs and PMs
t_1	End of probing [42]
t_2	End of constructing/configuration interval [42]
$t_s(v_i)$	The threat score of v_i [0,1]
$CoRes(v_i, v_j)$	The co-resident duration matrix between v_i and v_j
$rcr(v_i, v_j)$	Co-resident risk indicator matrix
$CoResFactor$	VM's co-resident rewards factor

Table 1.
Variable definition.

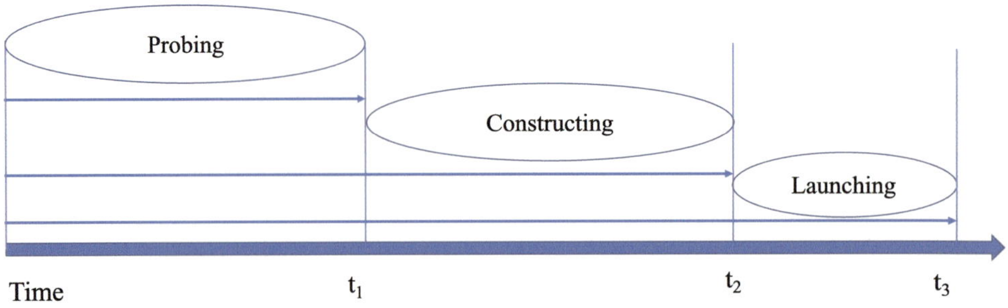

Figure 3.
The timeline of attacks [42].

The threat score, $t_s(v_i)$, reflects the potential risk to VM_i. It is a floating number between 0 and 1, with 0 representing no risk and 1 representing the highest risk. As illustrated in Eq. (1), the risk is also proportional to the co-resident duration time recorded in a matrix, $CoRes(v_i, v_j)$.

The VM's co-resident rewards factor, $CoResFactor$, is the crucial parameter in guiding RL training. It can be determined by where the co-resident attack cycle status of the VM resides. For example, the time of a co-existing VM pair on the same PM for a period of less than t_1 is considered to be safe. If there is a malicious VM, it means that it has not passed the probing stage yet. So, the risk of getting a co-resident attack is low. In this case, $\alpha_0 = 0$ is chosen. If the $CoRes(v_i, v_j)$ is between t_1 and t_2, the system needs to be aware that if a malicious VM exists in the pair; it reaches the constructing stage and moves closer to launching the attack. So, α_1 needs to be non-zero. While the VM pair co-exists on the same PM for more than t_2 time period, an attack could be launched. This is the situation to be avoided, so that α_2 is assigned to a more aggressive number.

3.3 Assumptions

Two assumptions guide the proposed model:

1. Co-resident risk is calculated among all the active VM pairs on the same PM, and each VM is randomly created with 1 to 15-time steps of the length.

2. All the jobs are batch jobs, and all the VMs might be malicious. The goal is to mitigate co-resident risk by avoiding the VM pairs' co-resident period from reaching t_2.

The system will be simulated under the above assumptions. The overall co-resident risk level, the number of co-resident attacks, and the VM slowdown ratio is the proposed system's evaluation metrics.

4. DRL-based VM scheduling system design and simulation

4.1 Mathematical background of RL and schematics

RL problems can be modeled as a Markov decision process (MDP) to find a policy by maximizing the accumulated rewards. An MDP has four tuples (S, A, P_a, R_a), where S is a set of states called state space, A is a set of actions called action space, P_a is the probability of state transition from s to s' under action a, and R_a is the immediate reward right after action a. There are two major methods to solve the reinforcement learning iteration problem. One is called $value - function$, and the other is $policy - gradient$. Q-Learning is an example of $value - function$, which has a function: $Q : S \times A \rightarrow R$. Before learning begins, Q is initialized to 0 or a base value. The core of the algorithm is a Bellman equation, which updates the Q value with new information:

$$Q^{new}(s_t, a_t) \leftarrow Q(s_t, a_t) + \alpha\left[r_t + \gamma \max_a Q(s_{t+1}, a) - Q(s_t, a)\right] \tag{3}$$

Here, α and γ are the learning rate and discount factor, respectively. r_t is the reward at the time step t. We adopt DeepRM [44] framework, which follows $policy - gradient$ with a deep neural network added into this system to solve large-scale RL tasks. This portion of the deep RL can be illustrated in **Figure 4**.

The nature of the $policy - gradient$ is to maximize the expected cumulative discount reward $E_{\pi\theta}\left[\sum_{t=0}^{\infty}\gamma^t r_t\right]$, which can be expressed as:

$$\nabla_\theta E_{\pi\theta}\left[\sum_{t=0}^{\infty}\gamma^t r_t\right] = E_{\pi\theta}\left[\nabla_\theta log\, \pi_\theta(s, a) Q^{\pi\theta}(s, a)\right] \tag{4}$$

Here, $\gamma \in (0, 1]$ is a discount factor for future rewards. r_t is the reward at the time step t. The VMM picks actions based on a $policy\ \pi : \pi(s, a) \rightarrow [0, 1]$, which is defined as the probability of action a taken in the state s. A manageable number of adjustable parameters, θ, are called the policy parameter. So, the policy can be represented as $\pi_\theta(s, a)$, and θ will be updated via gradient descent:

$$\theta \leftarrow \theta + \beta \sum_t \nabla_\theta log\, \pi_\theta(s_t, a_t) v_t \tag{5}$$

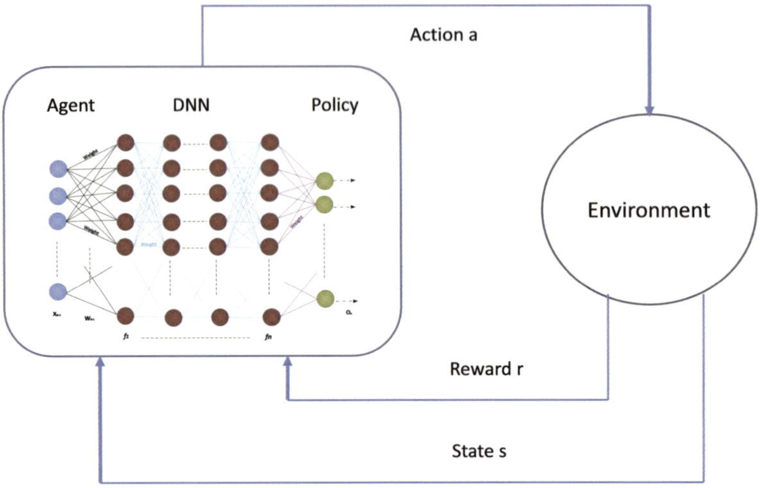

Figure 4.
Reinforcement learning with policy represented via DNN [44].

where β is the step size. The corresponding expected cumulative discounted reward $Q^{\pi\theta}(s, a)$ can be estimated by the empirically computed cumulative discounted reward v_t.

4.2 RL components design

Reinforcement learning is a unique type of machine learning paradigm, which has been successfully applied to task scheduling [45–49]. It contains several detailed components that need clarification. Here, we first define our state space, action space, and rewards before introducing the simulation system.

4.2.1 State space

RL is a model-free machine learning method; an agent learns from the trial-and-error process to interact with the environment. The state of the environment is defined as a vector of several components as shown in **Table 2**. They build the data structure of a *VM* which can be classified as:

1. Computing resources factors;

2. Security awareness factors (already introduced in **Table 1**).

The current allocation of the cluster resources can be retrieved by the mapping between VMs and resource slots available on the PM, which can be expressed as a matrix X.

4.2.2 Action space

It is assumed that VMs will be assigned to the PM if requested resources are available at each time step. The action space is defined by $\{0, 1, ..., n\}$, where 0 means

Vector component	Values	Data type
Computing Resources		
Hosting *PMs*	1	Integer
VM ID	1	Integer
Ideal length of finishing	1	Float
Resources requested	2	Integer
Start time	1	Float
Finish time	1	Float
$VM_{Delayed}$	1	Float
Security		
t_s	1	Float
CoRes	n	Integer
rcr	n	Float

Table 2.
Data structure of a VM.

no action taken, and 1 through n means to allocate a new VM on the $n's$ PM slot. After each action, the next state space is obtained by updating the recent mapping of X.

4.2.3 Rewards

A reward strategy is designed to guide the VM allocation agent toward our goal: Sufficiently utilize current resources to complete jobs on time and simultaneously minimize co-resident attacks. The reward function must be carefully designed to avoid contradiction. In the proposed system with a total of J active VMs, the rewards function consists of two terms:

1. VMs' completion factor, R_{VMD} (VM delay rewards), is defined by the accumulated $VM_{Delayed}$ from currently running VMs in the system. Here, T_j is the duration of the jth VM, v_j.

2. VMs' co-resident risk level, R_{RC} (runtime co-resident rewards), is defined by the co-resident risk indicator matrix $rcr(v_i, v_j)$.

They can be calculated accordingly as:

$$R_{VMD} = \sum_{j \in J} VM_{Delayed}(v_j) = \sum_{j \in J} \frac{-1}{T_j} \tag{6}$$

$$R_{RC} = \sum_{i,j \in J} (-1) \times rcr(v_i, v_j)/2 \tag{7}$$

The full rewards are calculated as a weighted sum of the two terms with weights ω_1 and ω_2. The overall rewards can be obtained by:

$$TotalRewards = \omega_1 \times R_{VMD} + \omega_2 \times R_{RC} \tag{8}$$

This rewards equation implies the objective of this novel VM allocation system is focused on:

1. Efficiently allocating VMs to minimize the VM delay time as shown in Eq. (6);

2. Aware of co-resident attacks through VM assignment correlations and take action to avoid the risk as shown in Eq. (7).

4.3 Simulation system design

4.3.1 System model

Similar to the DeepRM [44] framework, the proposed simulation system is illustrated in **Figure 5**. CPU and memory (MEM) are the two resources for limited constraint consideration. When the VM is assigned to the PM, a time step starts to count the duration of the VM. If there are other VMs simultaneously assigned to the same PM, the co-resident counter is also started to accumulate the time steps and recorded in $CoRes(v_i, v_j)$. VM requests arrive according to a Bernoulli process. The backlog queue houses all the incoming VMs waiting for allocation.

4.3.2 Co-resident duration matrix

The time steps will be recorded in the co-resident duration matrix as shown below. This small-scale example limits each CPU and MEM resource to five slots. Here, five VMs $\{VM_1, \ldots, VM_5\}$ are illustrated with the life cycles marked with a start and end

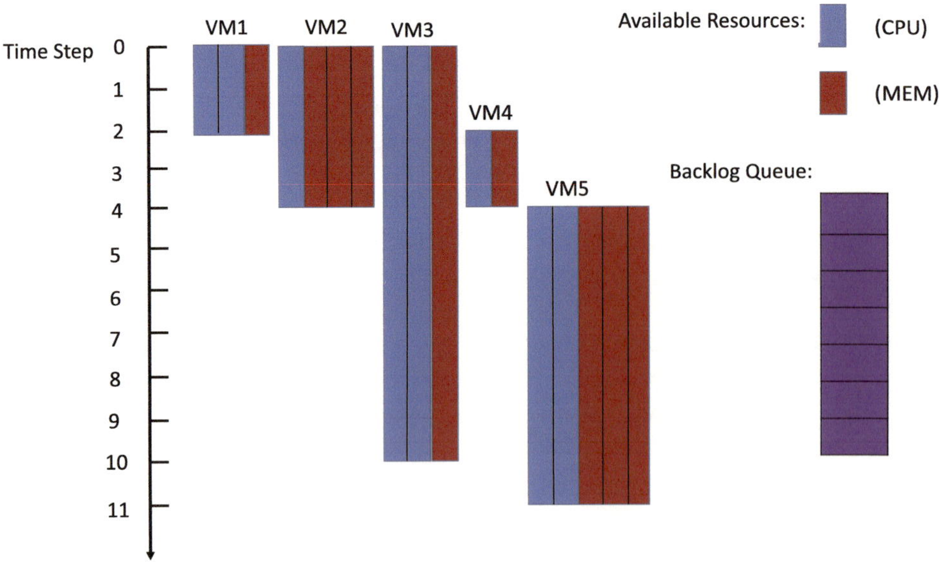

Figure 5.
Resource, time steps, job slots, and backlog queue in [44].

time steps as $VM_1 : [0, 2]$, $VM_2 : [0, 4]$, $VM_3 : [0, 10]$, $VM_4 : [2, 4]$, $VM_5 : [4, 11]$. Their life cycle lengths can be represented as $\{2, 4, 10, 2, 7\}$, respectively. The overlapped time steps shown in a co-resident duration matrix are:

$$CoRes(v_i, v_j) = \begin{bmatrix} 2 & 2 & 2 & 0 & 0 \\ 2 & 4 & 4 & 2 & 0 \\ 2 & 4 & 10 & 2 & 6 \\ 0 & 2 & 2 & 2 & 0 \\ 0 & 0 & 6 & 0 & 7 \end{bmatrix}$$

The proposed zero-trust strategy means all the VMs could be malicious, so the threat scores for all VMs are set to 1 ($t_s(v_i) = 1$). Thus, the co-resident risk indicator matrix is:

$$rcr(v_i, v_j) = 1 \times CoRes(v_i, v_j) \times 1 = CoRes(v_i, v_j) \tag{9}$$

4.3.3 Configuration interval

In the simulated system, five-time steps are chosen to represent t_1 and ten-time steps to represent the configuration interval t_2. In **Figure 5**, time is represented in a vertical direction. In order to mitigate the co-resident attacks, the system is designed to train the agent to avoid two VMs sharing the same PM for more than t_2 time interval. Based on the timeline of the attacks illustrated in **Figure 3**, different values will be assigned to the *CoResFactor* as shown in Eq. (2).

4.3.4 Risk mitigation strategies

When two VMs have overlapped time steps less than t_1, there is a minor risk of co-resident attacks, so $\alpha_0 = 0$; when two VMs have resided on the same PM for more than t_1 time steps, but less than t_2, co-resident attack risks start to accumulate. Thus, the first risk mitigation function is set to be: $\alpha_1 = k \times (t - t_1)$, while k has been chosen from $\{0, 0.25, 0.5, 1, 2\}$ to explore the efficiency of different choices. When two VMs have co-residence on the same PM for more than t_2 time steps, there is enough construction time for attacks to take place, so a more aggressive factor in the form of $k2$ is added to the reward function. The proposed system applies the second risk mitigation function: $\alpha_2 = k \times (t - t_1) + k2$, where $k2$ has been tested in the pool of $\{0, 1, 2, 3, \ldots\}$. A portion of the risk mitigation function design can be found in **Figure 6**, where all the k values are presented; only $k2 = 0$, $k2 = 1$, and $k2 = 2$ on top $k = 2$ are shown on the graph.

4.3.5 Software

The system is programmed in Python with the flowchart illustrated in **Figure 7**. First, the arriving VMs are placed in a backlog queue. If the queue is not empty, the scheduling system operates to find the optimized solution to assign VMs to PMs. At each time step, the system will update the co-resident duration matrix which reflects the current risk level and will guide the choice of risk mitigation strategies.

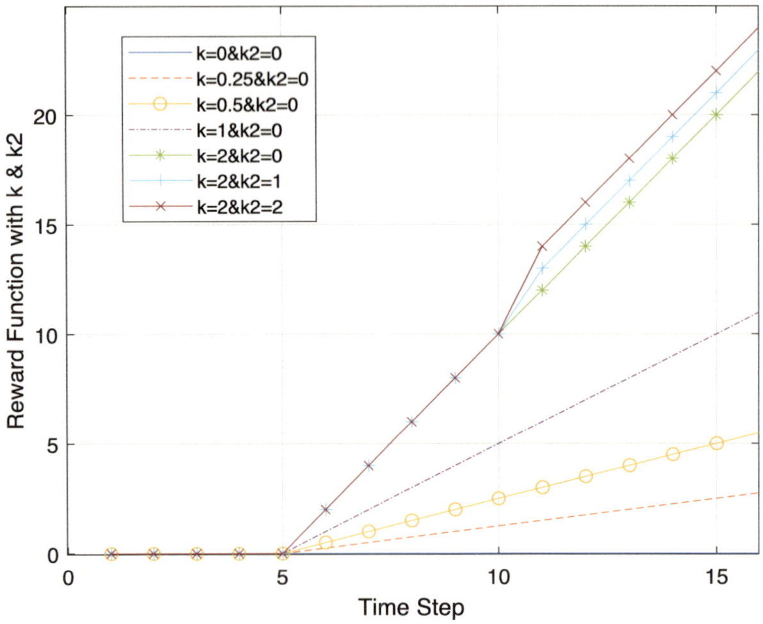

Figure 6.
Risk mitigation function design.

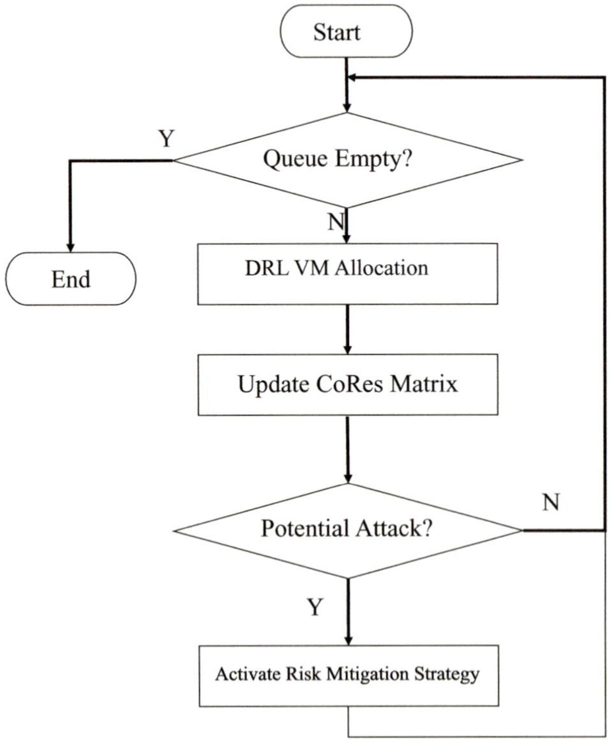

Figure 7.
The flowchart of the proposed system.

5. Simulation results analysis

The co-resident simulation program is built upon the Python-based DeepRM [44] open-source platform. The neural network is constructed by a fully connected hidden layer with 20 neurons, and a total of 89,451 parameters. Poisson distribution with a new arriving rate of 0.7 is chosen to simulate the VMs' dynamic arrival. All the results are obtained in 2500 iterations.

Our proposed system introduces co-resident risk mitigation to task scheduling by adding Eq. (7) to the total rewards calculation of Eq. (8). During the investigation, it was observed that the system performed differently, while manipulating the risk mitigation function parameters illustrated in **Figure 6**. The effectiveness of the proposed mitigation scheme can be analyzed by the RL rewards, VM slowdown ratio, and attack reduction.

5.1 Total rewards affected by risk mitigation factors

As illustrated in Eq. (8), total discounted rewards can be captured by taking both VM delay and co-resident risks into consideration. Since there is no preference between the two, $\omega_1$and ω_2 in Eq. (8) are both set to 1. In the first experiment, $k2$ is set to 0, and k is chosen from 0, 1, and 2. The recorded total accumulated rewards in **Figure 8** explain that a smaller k value leads to a larger reward (Note: the reward is negative). When $k = 0$there is no risk mitigation. As k increases, more mitigation influence will be placed in the system, and the total discounted reward decreases.

DeepRM provides DRL and other heuristics VM allocation methods, such as tetris, random allocation, and small jobs first (SJF), for comparison. Although those methods do not have co-resident risk mitigation features, the total discounted rewards can illustrate how severe the cybersecurity risks they are experiencing. Two user cases are

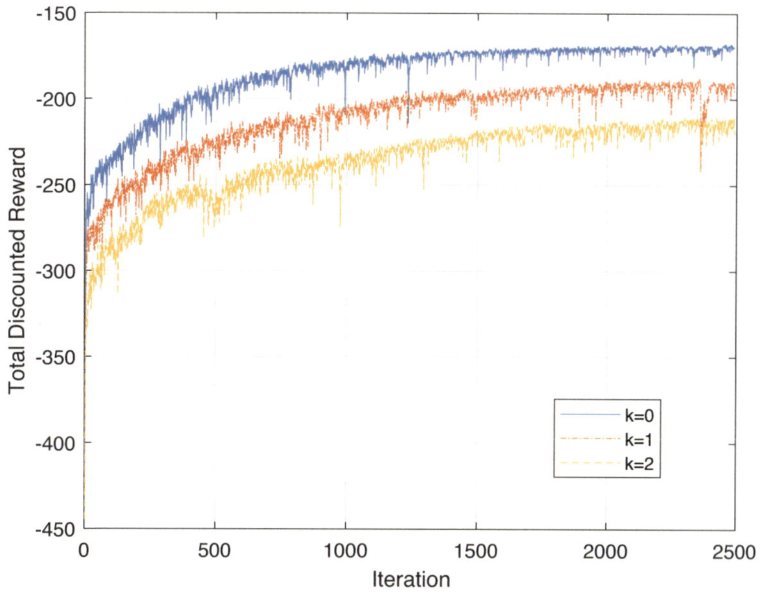

Figure 8.
The total rewards accumulated from different k values.

Approved for Public Release on 01 June 2022; Distribution Unlimited; Case number: AFRL-2022-2581.

Methods	Tetris	Random	SJF	DRL
Case 1	−644.94	−426.81	−501.40	−320.17
Case 2	−267.88	−167.50	−187.33	−142.69

Table 3.
Total discounted rewards.

shown in **Table 3**. A negative number with a larger absolute value means a worse situation.

5.2 Slowdown ratio affected by risk mitigation factors

The metric to measure the efficiency of VM scheduling is to calculate the $VM_{Delayed}$ as defined in **Table 2**. In programming, the slowdown ratio is utilized. Each VM has its own expected life cycle shown as the "Ideal length of finishing" in **Table 2**. It also has a length marked by time step when generated. When the VM is assigned to a PM, the "Start time" is marked. At the time of finishing, a "Finish time" is recorded. The parameter *Slowdown* is calculated by Eq. (10). Ideally, if there is no delay in the execution, $Slowdown = 1$, but in the actual application, many factors can cause the delay. Thus, $Slowdown \geq 1$.

$$Slowdown = (FinishTime - StartTime)/VMLength \tag{10}$$

With an increment of k value, more rewards are generated to mitigate the potential co-resident risks through the risk mitigation function. As a matter of fact, it sacrifices the VM completion time, so the slowdown ratio increases. Experiments show that "Random" allocation of VMs has the largest average slowdown ratio. If using "Random" slowdown ratio as a baseline, the percentage of slowdown ratio reduction from the baseline data is shown in **Table 4**.

5.3 Co-resident attacks reduction by risk mitigation factors

Considering the goal of mitigating co-resident attacks, a group of experiments is conducted to represent the effectiveness of different risk mitigation function parameters under RL scenario. **Figure 9** illustrates the total counts of co-resident attacks if k and $k2$ are set as in **Figure 6**. If $k = 2$ and $k2 = 1$, the count reduces dramatically compared with $k = 0$ and $k2 = 0$, where there is no mitigation applied.

6. Conclusions and future work

This chapter addresses the importance of cybersecurity awareness in cloud computing resource management. The proposed RL-based scheduling method takes both

Methods	Tetris	SJF	DRL ($k = 0$)	DRL ($k = 1$)	DRL ($k = 2$)
ASR	63%	60%	72%	71%	68%

Table 4.
VM slowdown ratio over random method.

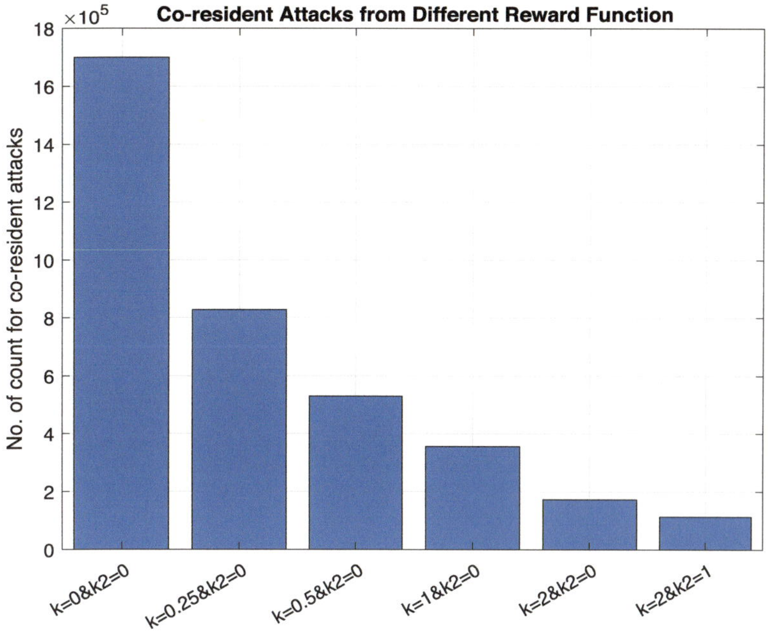

Figure 9.
The potential co-resident attacks by different k and k2 selection strategies.

VM slowdown time and co-resident attack risks mitigation into consideration. The co-resident risk model under no-trust conditions is formed. As a result, the problem is narrowed down to minimizing the co-tenancy on the same *PM* among all the active VMs. Finally, a DRL-based task scheduling system is simulated with proposed risk mitigation factors.

This chapter proves that much can be explored in resource management and risk mitigation in cloud computing. It is evident that ML obtained much attention recently, and more applications are being developed in this direction. Although there is a concern about training costs under a deep learning algorithm, it outperforms other methods in adaptation to a more dynamic environment, which makes it outstanding. If designed properly, the computational burden can be shifted to offline. The above experiment results are obtained by using MacBook Air with a 2.2GHz dual-core Intel i7 processor and 8GB memory. It takes 3 minutes per 2500 iterations to train the policies. While applying the pre-trained model to take actions during runtime testing, it will not take longer than 2 seconds for the longest allocation decision. The results show applying reinforcement learning to co-resident risk mitigation is plausible. Different mitigation strategies lead to different VM completion ratios and risk levels. The proposed strategies proved to be helpful in searching for VM allocation improvement with consideration of both VM completion constraints and co-resident risk awareness. In the future, a more in-depth investigation of the reward equation design will be conducted. A thorough search accompanied by mathematical models to discuss the convergence will be explored. An advanced cost function will be developed with resources and security constraints. Multi-agent reinforcement learning will be applied to extend the model of this research and the efficiency will be tested and compared.

Acknowledgements

This work was supported in part by the Air Force Research Laboratory and Department of Education MSEIP grant award no. P120A180114, Texas A & M Engineering Experiment Station Annual Research Conference Project (TEES TARC) Award 28-235980-00020, and the National Science Foundation grant award no. OAC 1827243. The views and conclusions contained herein are those of the authors and should not be interpreted as necessarily representing the official policies or endorsements, either expressed or implied, of the Air Force Research Laboratory or the U.S. Government.

Conflict of interest

The authors declare no conflict of interest.

Author details

Suxia Cui[1*†] and Soamar Homsi[2†]

1 Prairie View A&M University, Prairie View, TX, USA

2 US Air Force Research Laboratory, Rome, NY, USA

*Address all correspondence to: sucui@pvamu.edu

† These authors contributed equally.

References

[1] Hossain S. Chapter 1 Cloud Computing Terms, Definitions, and Taxonomy. In: Cloud Computing Service and Deployment Models: Layers and Management. IGI Global. 2013. pp. 1-25

[2] BarbosaAndrea FP, Charão AS. Impact of pay-as-you-go cloud platforms on software pricing and development: A review and case study. Computational Science and Its Applications. 2012;**7336**: 404-417

[3] Smith JE, Nair R. Virtual Machines: Versatile Platforms for Systems and Processes. Amsterdam, Netherlands: Elsevier; 2005

[4] Tank D, Aggarwal A, Chaubey N. Virtualization vulnerabilities, security issues, and solutions: a critical study and comparison. International Journal of Information Technology. 2022;**14**:847–862

[5] Hasan MM, Rahman MA. Protection by Detection: A Signaling Game Approach to Mitigate Co-Resident Attacks in Cloud. In: 2017 IEEE 10th International Conference on Cloud Computing (CLOUD). Honolulu, HI, USA; 2017. pp. 552-559

[6] Ding J, Sha L, Chen X. Modeling and evaluating IaaS cloud using performance evaluation process algebra. In: 2016 22nd Asia-Pacific Conference on Communications (APCC). Yogyakarta, Indonesia. 2016. pp. 243-247

[7] Addya SK, Turuk AK, Satpathy A, Sahoo B, Sarkar M. A strategy for live migration of virtual machines in a cloud federation. IEEE Systems Journal. 2019; **13**(3):2877-2887

[8] Velayudhan Kumar MR, Raghunathan S. Heterogeneity and thermal aware adaptive heuristics for energy efficient consolidation of virtual machines in infrastructure clouds. Journal of Computer and System Sciences. 2016;**82**(2):191-212

[9] Mthunzi SN, Benkhelifa E, Alsmirat MA, Jararweh Y. Analysis of VM communication for VM-based cloud security systems. In: 2018 Fifth International Conference on Software Defined Systems (SDS). 2018. pp. 182-188

[10] Sane BO, Niang I, Fall D. A review of virtualization, hypervisor and VM allocation security: Threats, vulnerabilities, and countermeasures. In: 2018 International Conference on Computational Science and Computational Intelligence (CSCI). Las Vegas, NV, USA. 2018. pp. 1317-1322

[11] Navamani BA, Yue C, Zhou X. Discover and Secure (DaS): An Automated Virtual Machine Security Management Framework. In: 2018 IEEE 37th International Performance Computing and Communications Conference (IPCCC). Orlando, FL, USA. 2018. pp. 1-6

[12] Kong T, Wang L, Ma D, Xu Z, Yang Q, Chen K. A Secure Container Deployment Strategy by Genetic Algorithm to Defend against Co-Resident Attacks in Cloud Computing. In: 2019 IEEE 21st International Conference on High Performance Computing and Communications; IEEE 17th International Conference on Smart City; IEEE 5th International Conference on Data Science and Systems (HPCC/SmartCity/DSS). Zhangjiajie, China: IEEE; 2019. pp. 1825-1832

[13] Qiao A, Choe SK, Subramanya SJ, Neiswanger W, Ho Q, Zhang H, et al. Pollux: Co-adaptive cluster scheduling

for goodput-optimized deep learning. In: 15th USENIX Symposium on Operating Systems Design and Implementation (OSDI 21). Virtual Conference; 2021. pp. 1-18

[14] Xiao W, Ren S, Li Y, Zhang Y, Hou P, Li Z, et al. AntMan: Dynamic scaling on GPU clusters for deep learning. In: 14th USENIX Symposium on Operating Systems Design and Implementation (OSDI 20). virtual conference; 2020. pp. 533-548

[15] Gu J, Chowdhury M, Shin KG, Zhu Y, Jeon M, Qian J, et al. Tiresias: A GPU cluster manager for distributed deep learning. In: 16th USENIX Symposium on Networked Systems Design and Implementation (NSDI 19). Boston, MA: USENIX Association; 2019. pp. 485-500

[16] Zhao J, Rodríguez MA, Buyya R. A deep reinforcement learning approach to resource management in hybrid clouds harnessing renewable energy and task scheduling. In: 2021 IEEE 14th International Conference on Cloud Computing (CLOUD). Chicago, IL, USA; 2021. pp. 240-249

[17] Gupta K, Katiyar V. Survey of resource provisioning heuristics in cloud and their parameters. International Journal of Computational Intelligence Research. 2017;**13**(5):1283-1300

[18] Gawali MB, Shinde SK. Task scheduling and resource allocation in cloud computing using a heuristic approach. Journal of Cloud Computing. 2018;**7**(1):4

[19] Dorigo M, Birattari M, Stutzle T. Ant colony optimization. IEEE Computational Intelligence Magazine. 2006;**1**(4):28-39

[20] Qin Y, Wang H, Zhu F, Zhai L. A multi-objective ant colony system algorithm for virtual machine placement in traffic intense data centers. IEEE Access. 2018;**6**:58912-58923

[21] Tawfeek MA, El-Sisi A, Keshk AE, Torkey FA. Cloud task scheduling based on ant colony optimization. In: 2013 8th International Conference on Computer Engineering Systems (ICCES). Cairo, Egypt; 2013. pp. 64-69

[22] Patel KD, Bhalodia TM. An efficient dynamic load balancing algorithm for virtual machine in cloud computing. In: 2019 International Conference on Intelligent Computing and Control Systems (ICCS). Madurai, India; 2019. pp. 145-150

[23] Jia H, Liu X, Di X, Qi H, Cong L, Li J, et al. Security strategy for virtual machine allocation in cloud computing. Procedia Computer Science. 2019;**147**: 140-144

[24] Miao F, Wang L, Wu Z. A VM placement based approach to proactively mitigate co-resident attacks in cloud. In: 2018 IEEE Symposium on Computers and Communications. Natal, Brazil: IEEE; 2018. pp. 00285-00291

[25] Han Y, Chan J, Alpcan T, Leckie C. Virtual machine allocation policies against co-resident attacks in cloud computing. In: 2014 IEEE International Conference on Communications (ICC). Sydney, NSW, Australia: IEEE. 2014. pp. 786-792

[26] Yang J, Jiang B, Lv Z, Choo KKR. A task scheduling algorithm considering game theory designed for energy management in cloud computing. Future Generation Computer Systems. 2020; **105**:985-992

[27] Patra MK, Sahoo S, Sahoo B, Turuk AK. Game theoretic approach for real-time task scheduling in cloud

computing environment. In: 2019 International Conference on Information Technology (ICIT). Bhubaneswar, India: IEEE; 2019. pp. 454-459

[28] Han K, Cai X, Rong H. An evolutionary game theoretic approach for efficient virtual machine deployment in green cloud. In: 2015 International Conference on Computer Science and Mechanical Automation (CSMA). Hangzhou, China; 2015. pp. 1-4

[29] Narwal P, Singh SN, Kumar D. Predicting strategic behavior using game theory for secure virtual machine allocation in cloud. In: Networking Communication and Data Knowledge Engineering. Singapore: Springer; 2018. pp. 83-92

[30] Narwal P, Kumar D, Singh SN. A hidden markov model combined with Markov Games for intrusion detection in cloud. Journal of Cases on Information Technology (JCIT). 2019;**21**(4):14-26

[31] Han Y, Alpcan T, Chan J, Leckie C. Security games for virtual machine allocation in cloud computing. In: International Conference on Decision and Game Theory for Security. Fort Worth, TX, USA: Springer; 2013. pp. 99-118

[32] Pahlevan A, Qu X, Zapater M, Atienza D. Integrating heuristic and machine-learning methods for efficient virtual machine allocation in data centers. IEEE Transactions on Computer-Aided Design of Integrated Circuits and Systems. 2017;**37**(8): 1667-1680

[33] Witanto JN, Lim H, Atiquzzaman M. Adaptive selection of dynamic VM consolidation algorithm using neural network for cloud resource management. Future Generation Computer Systems. 2018;**87**:35-42

[34] Zhang J, Xie N, Zhang X, Yue K, Li W, Kumar D. Machine learning based resource allocation of cloud computing in auction. Comput Mater Continua. 2018;**56**(1):123-135

[35] Liu Z, Zhang H, Rao B, Wang L. A reinforcement learning based resource management approach for time-critical workloads in distributed computing environment. In: 2018 IEEE International Conference on Big Data (Big Data). Seattle, WA, USA: IEEE; 2018. pp. 252-261

[36] Joseph L, Mukesh R. To detect malware attacks for an autonomic self-heal approach of virtual machines in cloud computing. In: Fifth International Conference on Science Technology Engineering and Mathematics (ICONSTEM). Chennai, India: IEEE; 2019. pp. 220-231

[37] Abazari F, Analoui M, Takabi H. Multi-objective response to co-resident attacks in cloud environment. International Journal of Information and Communication Technology Research. 2017;**9**(3):25-36

[38] Liu Y, Ruan X, Cai S, Li R, He H. An optimized VM allocation strategy to make a secure and energy-efficient cloud against co-residence attack. In: International Conference on Computing, Networking and Communications (ICNC). Maui, HI, USA: IEEE; 2018. pp. 349-353

[39] Berrima M, Nasr AK, Ben RN. Co-location resistant strategy with full resources optimization. In: Proceedings of the 2016 ACM on Cloud Computing Security Workshop. Hofburg Palace, Vienna, Austria; 2016. pp. 3-10

[40] Levitin G, Xing L, Dai Y. Co-residence based data vulnerability vs. security in cloud computing system with

random server assignment. European Journal of Operational Research. 2018; **267**(2):676-686

[41] Xing L, Levitin G. Balancing theft and corruption threats by data partition in cloud system with independent server protection. Reliability Engineering & System Safety. 2017;**167**:248-254

[42] Zhang Y, Li M, Bai K, Yu M, Zang W. Incentive compatible moving target defense against vm-colocation attacks in clouds. In: IFIP International Information Security Conference. Heraklion, Crete, Greece: Springer; 2012. pp. 388-399

[43] Wang X, Wang L, Miao F, Yang J. SVMDF: A secure virtual machine deployment framework to mitigate co-resident threat in cloud. In: 2019 IEEE Symposium on Computers and Communications (ISCC). Barcelona, Spain; 2019. pp. 1-7

[44] Miao H, Alizadeh M, Menache I, Kandula S. Resource management with deep reinforcement learning. In: HotNet '16: Proceedings of the 15th ACM Workshop on Hot Topics in Networks. Atlanta, Georgia, USA. 2016. pp. 50-56

[45] Mao H, Schwarzkopf M, Venkatakrishnan SB, Meng Z, Alizadeh M. Learning Scheduling Algorithms for Data Processing Clusters. In: Proceedings of the 2019 ACM Special Interest Group on Data Communication (SIGCOMM). Beijing, China; 2019. p. 270–288

[46] Tuli S, Ilager S, Ramamohanarao K, Buyya R. Dynamic scheduling for stochastic edge-cloud computing environments using A3C learning and residual recurrent neural networks. In: IEEE Transactions on Mobile Computing. 2022 March;**21**(3):940-954

[47] Tuli S, Poojara SR, Srirama SN, Casale G, Jennings NR. COSCO: Container Orchestration Using Co-Simulation and Gradient Based Optimization for Fog Computing Environments. IEEE Transactions on Parallel and Distributed Systems. 2022;**33**(1):101-116

[48] Paeng B, Park IB, Park J. Deep reinforcement learning for minimizing tardiness in parallel machine scheduling with sequence dependent family setups. IEEE Access. 2021;**9**(10):1390-1401

[49] Asheralieva A, Niyato D, Xiong Z. Auction-and-learning based lagrange coded computing model for privacy-preserving, secure, and resilient mobile edge computing. In: IEEE Transactions on Mobile Computing. 2021;early access. pp. 1-2

Section 2

Applied Intelligence

Chapter 5

Velocity Planning via Model-Based Reinforcement Learning: Demonstrating Results on PILCO for One-Dimensional Linear Motion with Bounded Acceleration

Hsuan-Cheng Liao, Han-Jung Chou and Jing-Sin Liu

Abstract

The time-optimal control problem (TOCP) has faced new practical challenges, such as those from the deployment of agile autonomous vehicles in diverse uncertain operating conditions without accurate system calibration. In this study to meet a need to generate feasible speed profiles in the face of uncertainty, we exploit and implement probabilistic inference for learning control (PILCO), an existing sample-efficient model-based reinforcement learning (MBRL) framework for policy search, to a case study of TOCP for a vehicle that was modeled as a constant input-constrained double integrator with uncertain inertia subject to uncertain viscous friction. Our approach integrates learning, planning, and control to construct a generalizable approach that requires minimal assumptions (especially regarding external disturbances and the parametric dynamics model of the system) for solving TOCP approximately as the perturbed solutions close to time-optimality. Within PILCO, a Gaussian Radial basis functions is implemented to generate control-constrained rest-to-rest near time-optimal vehicle motion on a linear track from scratch with data-efficiency in a direct way. We briefly introduce the importance of the applications of PILCO and discuss the learning results that PILCO would actually converge to the analytical solution in this TOCP. Furthermore, we execute a simulation and a sim2real experiment to validate the suitability of PILCO for TOCP by comparing with the analytical solution.

Keywords: model-based reinforcement learning (MBRL), applied reinforcement learning, time-optimal control problem (TOCP), velocity learning, vehicle control

1. Introduction

Optimal control–based approaches have played key roles in trajectory planning or replanning and in optimization of control inputs with numerous applications, such as robotics and autonomous driving. These approaches have recently been used in autonomous systems. Optimal control formulations typically require accurate knowledge of the dynamics and can account for a more general set of constraints and objectives (performance measures) [1, 2] relative to other approaches. Many cost functions, such as those for time and energy, are used to define the desired behavior of the controlled system. As part of an effort to enhance working efficiency and productivity by having tasks be completed as fast as possible, especially for tasks involving repetitive state-to-state transfer in trajectory execution, Time-Optimal Control Problem (TOCP), for which the objective function is the terminal time, has been extensively studied. For TOCP, control bounds, different boundary conditions or paths with different curvature profiles and lengths, and various choices of the physical parameters such as mass, friction, can produce different velocity solutions. As a result, different maximum velocities and different travel times are produced. It was studied first on articulated robot manipulators for industrial and aerospace applications [3–7].

In recent years, the deployment of agile autonomous systems such as autonomous driving vehicles [8, 9], mobile robots [1, 10, 11] and new robot platforms such as humanoid robots [12] and unmanned aerial vehicles (uavs) have posed new increasingly important challenges to TOCP. Furthermore, a crucial aspect of traditionally solving TOCP is that analytical or learned time-optimal velocity solution are both platform and path (task) dependent, therefore, model-based and goal-directed. These challenges lie in several aspects including nonlinear, nonconvex, multi-dimensional state and control spaces, as well as various platform dependent constraints, and most importantly uncertainties of the environment in real world problems and make the model only able to approximate the reality. Therefore, computational solutions to TOCP based on an not so accurate model are not reliable and not practical for online applications. By contrast, learning-based control algorithms for dynamical systems learn to generate the desired system behavior without any complicated system formalism or predefined controller parameters a priori and thereby achieve more generalization and platform independency. One promising approach in the context of intelligent planning and control involves the use of reinforcement learning (RL) [13, 14] for learned behaviors, which can be viewed as a class of optimal control resolutions. The effectiveness and performance of RL is of task instance and platform specific, i.e. as a function of the transition and reward functions induced by the evaluated policy and the system dynamics. Therefore, to address the practical challenges in facilitating RL algorithms for a wide range of real-world decision-making problems such as the autonomous vehicles in the context of diverse driving scenarios, it is generally believed that only by proposing specific applications of RL on concrete cases can better demonstrate related issues and in which algorithm works well for a specific task instance [15].

With this aim for study, in this paper, the learning goal is to recover a near time-optimal rest-to-rest one-dimensional linear motion on a double integrator with embedded uncertainties (frictions of motion) and constant control constraint. We assume no prior knowledge of any parametric system dynamics model for deriving the optimality conditions. In addition, the characteristics of the learning task is that the vehicle mass is uncertain and the environment characteristics such as friction is unknown, and both parameters affect the maximum speed at each position along the path. It is worth mentioning that any single-input controllable second-order system is feedback equivalent to a double integrator. Thus, we use a simplified but still general, precise enough vehicle model, damped double integrator, as the foundation and demonstration for TOCP for more complicated, high-dimensional nonlinear vehicle model. The analytical solution of TOCP to double integrator subject to constant acceleration bound is known of a second-order ODE solution with an assumption by bang-bang control.

In the scope of this paper, our MBRL-based approach to recover the time-optimal motion of double integrator features a two-stage process for integrating learning, planning and control. In the first stage, we employ a Model-Based Reinforcement Learning (MBRL) framework, Probabilistic Inference for Learning Control (PILCO) [16], to generate a control-constrained rest-to-rest near time-optimal motion from scratch. The analytical solution is used as a baseline for effectively assessing the learning results. Because Monte-Carlo updates of the parametrized policy renders it difficult to incorporate velocity limits within an instance of locomotion, hence, as a second stage, we apply rescaling to give a speed profile respecting the additional velocity limit.

The outcome is then a time-optimal velocity profile under both velocity and control constraints, with no prior required knowledge and replanning. Both simulation and sim-to-real experiments are conducted and confirm our approach applicable.

Our main contributions include:

1. A novel application of a model-based reinforcement learning algorithm, PILCO, serving as adaptive optimal control, learns from scratch the time-optimal control of double-integrator vehicle model in the presence of uncertainties. The near time-optimality and data-efficiency is observed in light of the simulation and sim2real experimental validation we present.

2. A distinct feature of this work compared to the majority of the related literature [17–19] is that the learning results are evaluated and verified effectively by the analytical time-optimal motion, instead of based on a lot of test scenarios. The results, which are valid for line following of a single input controllable second-order uncertain system, allow interpretations in terms of integrated learning, planning, and control.

The remainder of the article is structured as follows. In Section 2, we summarize the approaches to the TOCP, specifically conventional and RL approaches. In Section 3, we outline the key elements of the PILCO algorithm, which are dynamics modeling, trajectory prediction, policy evaluation, and policy improvement. In Section 4, we present our simulation results on time-optimal velocity learning for an autonomous vehicle with double integrator dynamics whose analytical solution to TOCP is derived in Appendix A as the verification baseline. A sim2real experimental validation on a low-cost car along with discussions in Section 5 is provided. Finally, Section 6 provides conclusions.

2. Related work

The aim of time-optimal vehicle control is to control a vehicle such that it reaches a target state as quickly as possible (e.g., in racing or emergencies). Minimal-time velocity profile along a prespecified curve, as a subclass of TOCP subject to hard control constraints resulting from input saturation, state constraints and external disturbance, is nowadays applied to a variety of modern autonomous systems such as autonomous driving, uav and robotics. The existence of time-optimal trajectories is guaranteed by Pontryagin Maximum Principle (PMP) [20] for a vehicle with explicit dynamics model in state-space form [21]. The derivations of the optimal control and state trajectories are generally computationally expensive; computationally cheaper yet accurate methods are required, especially considering the need for rapid (even real-time) computation in most real-world industrial or engineering systems in response to changes in operating conditions and the environment. In this section, we briefly present the most common numerical approaches for systems with known dynamics model developed in robotics and autonomous vehicles; we then discuss RL-based approaches that handle uncertainty.

2.1 Approaches to the TOCP with known dynamics

Solutions to the TOCP can be categorized as complete or decoupled approaches. In the complete approach, where the aim is to solve challenging problems with general vehicle dynamics and constraints in their entirety, the optimal state and input trajectories are simultaneously determined; some direct or indirect transcription methods have been developed as part of this approach for trajectory optimization [2, 22, 23] and played an important role in their numerical performance of trajectories. By contrast, in the decoupled approach (e.g., the path velocity decomposition approach and path-constrained trajectory generation approach), the trajectory generation is decomposed into two subproblems for the path geometry to be decoupled from the velocity along the geometric path. In the decoupled approach, the first step is planning a geometric path for connecting two states (configurations or poses) in adherence to geometric constraints, such as obstacle avoidance or smoothness requirements, and the second step is designing a time-scaling function (that represents either information on timing or the velocity and acceleration) along the planned state-to-state-transfer geometric path. This approach results in a one-parameter family of velocity profiles, i.e. the parametrization of the vehicle-position-dependent velocity along the path as a function of the single path parameter of the arc length. The velocity and acceleration of the vehicle on each position of the path can be altered by the design of the time-scaling function respecting smoothness requirement (such as small jerk) and fixed boundary conditions (such as the initial and target positions and velocities are precisely specified) and the kinodynamic constraints constraints. A fair amount of literature on maximizing the speeds along the path with the acceleration, torque, jerk (or torque/acceleration derivative) constraints [3, 8, 10, 11]. In general, a model predictive control (MPC) framework can be used in the decoupled approach to generate the safe velocity profile and the input commands for following a given planned geometric path in terms of known system dynamics [24]. The following three methods have been commonly used for TOCP.

2.1.1 Hamilton-Jacobi-bellman (HJB) equation

A popular approach to obtain time-optimal motion for a system with known dynamic model and fixed boundary conditions under the safety and kinodynamic constraints of a vehicle is via optimal control or model predictive control formulations. This approach requires the derivation of optimality conditions for the state trajectories and control policies based on PMP or Dynamic Programming Principle (DPP) [20]. This yields the Two-point Boundary Value Problem of HJB partial differential equations with initial condition on the state and final condition on the costate for time-optimization of trajectories. The advantage of generality is that more general state and input constraints and objective functions can be taken into account at the cost of heavy numerical burden. For example, time-optimality can be traded off against energy to yield less aggressive control to steer the vehicles slower but smoother. Additionally, HJB equation approach is a practically useful approach in that many numerical solvers of HJB equations are available.

2.1.2 Convex optimization (CO)

The Hamiltonian of TOCP for robotic manipulator is shown to be convex with respect to the control input. TOCP is transformed into a convex optimization problem with a single state through a nonlinear change of variables [4], where the acceleration and velocity at discretized locations on the path are the optimization variables. Then, followed by [5] the work is further extended to meet speed dependent requirements. Such approach is simple and robust thanks to the existing convex optimization libraries, yet only convex objective functions can be concerned. However, the convex optimization program contains a large number of variables and inequality constraints, making it slow and less suitable for real-time applications.

2.1.3 Numerical integration (NI)

Since the vehicle velocity highly depends on the path to be followed, the universally applicable decoupled approach splits the motion planning problems to two sub-problems of finding a geometric path and planning the velocity at each position of the vehicle along the path to manage the computational complexity to generate a suboptimal motion trajectory. This result in a one-parameter family of velocity profiles, or the velocity (bound) along the path depending on the vehicle position on the path is parametrized as a function of the single path parameter (or a scalar curvilinear abscissa coordinate) s, usually the arc length. By description of the dynamics and constraints along the path to be followed on the $(s, \dot{s})$ phase plane, then this method generates the velocity limit curve on the phase plane from the velocity and acceleration bounds. The travel time is determined by the path velocity $\dot{s}$ along the path or the time scaling function $s(t)$ that is solved by optimization tools to meet the imposed constraints. The generation of minimum-time velocity profile along the given path is greatly simplified to the determination of switching structure in the phase plane [12]. Essentially, NI searches for switching points on the phase plane and establishes the velocity profile by integrating forward with the acceleration limits and backward with the deceleration limits pivoting from those points.

2.2 Reinforcement learning (RL)

RL refers to the learning of a policy, defined as a mapping from state space to action space, by means of maximizing a reward the agent receives from the environments it operates and interacts. When the system dynamics is unknown, with the dynamical system modeled as a reward-maximizing RL agent and the desired behavior expressed as a reward function, the system can be trained to automatically execute an optimal sequence of actions (trajectories) under the present environmental conditions to complete a given mission. The RL framework reformulates the problem as a Markov Decision Process for the autonomous agent, which maximizes the long-term rewards and does not necessarily need the transition dynamics beforehand. RL offers a diverse set of model-based and model-free algorithms to improve the performance of

RL agent based on the reward it received. The mission may involve integrated planning and control with time and computational resource budget in real-time applications, a system model therefore is a good informative basis for predicting the behaviors and enhancing performance, instead of tuning the behaviors frequently and manually in practical situations. However, it is often infeasible to derive a precise, analytical model that is provably correct for the actual system within regions and time horizon that the model is valid, since the existence of parametric uncertainties, unmodelled dynamics, external disturbance in perception and agent-environment interaction is inevitable. To ameliorate the problems from these interwoven factors that affect planning and control performance, researchers in the field of modeling and control have been increasingly interested in model-building or model learning in a data-driven setting based on nonparametric and probabilistic models [25, 26] as a means to enhance the performance of the underlying controlled system. Model-based RL (MBRL), complementary to the control approaches such as robust and adaptive control, model predictive control and fuzzy control, is an attractive intelligent model-based control approach that integrates learning, planning and control: it learns a dynamics model and then the derived characteristics of a learned model is exploited for generating trajectories and learning the policy. While the model-free RL attracts the most scientific interest, in MBRL algorithms that employ derived or learned system dynamic models, during policy evaluation, the state evolution calculated by a predictive model under a given policy can be used to estimate the impact of the policy on the reward. Therefore MBRL is more data-efficient in the context of diverse goal-directed planning tasks since fewer interactions between the agent and the environment are required to learn a good policy faster, in contrast to some model-free approaches. A surging number of researches demonstrated that learning, planning and control of autonomous systems such as robotics and self-driving vehicles can be cast as MBRL tasks, due to the use of an accurate, reliable learned model of the agent-environment interactions as an internal simulation in task execution and the basis for any optimization and real time control for achieving highly effective control performance [13, 14, 27]. Despite its faster convergence over the model-free frameworks, MBRL suffers from model bias and accumulated model prediction errors that greatly affect the control policy learning and rewards by leveraging the model-generated trajectories characteristics. To improve model learning performance and thus policy learning performance (the controller parameters are learned in light of currently learned model), system transition modeling or model fitting techniques ranging from deterministic methods such as physics-based (first principles based) formulation, to stochastic methods are developed [15]. Among which, nonparametric regression models such as Gaussian Process (GP) that extracts the information from the sampled data with the high data efficiency to make accurate predictions in PILCO [16]. In contrast to other probabilistic models that maintain a distribution over random variables, GP builds one over the underlying functions that generate the data. Therefore, it has no prior assumption on the function mapping current states and actions to future states. The flexibility and expressiveness GP to refine the uncertainty estimate offers makes it an effective approximator for modeling the unknown system dynamics (transition function from input data to output observation or measurement) that continuously evolves over time (i.e. trajectories), and is employed in this study. Some popular GP implementations are tabulated in [28] for practitioners. The most recent open source MBRL-Lib is released to reduce the coding burden [29].

3. MBRL for time-optimal vehicle motion

The imperfect modeling of system dynamics and perception of environment with significant noise makes machine learning a viable approach to the practical, near minimum-time velocity planning for autonomous systems that do not rely on heavy dynamic model-based computations using identification techniques. In order to find the time-optimal control policy for a vehicle dynamics model with uncertainties along a predefined path, in RL setting, it is learned from limited trial driving experiences the action (sequence) to be applied at each possible state confined on the path for the uncertain system (i.e. system dynamics along the path) with the received rewards as feedback, aiming to allocate higher trajectory reward, to determine the next observed state. There are a number of choices of RL algorithms. An earlier work [17] used Q-learning for car-like vehicle motion planning. Another study [18] analyzed transfer learning in obstacle avoidance behaviors in similar environments with similar obstacle patterns, where the state of the environment is represented by the obstacle pattern. Recently, a model-free actor-critic RL algorithm was applied to time-optimal velocity planning along an arbitrary path in [19]. That study demonstrated that the incorporation of velocity computation through the exploitation of a vehicle dynamics model is practically feasible for improving learning outcomes. These studies exemplify the practical utility of RL in improving a model's ability to control vehicle driving (encoded as a set of trajectories). These work, among others, shows that vehicle driving skill (encoded as a set of trajectories) learning via RL is promising for real autonomous driving. Along this line of work, in the present study, the learning task for vehicle control demonstrated is a one-dimensional vehicle maneuvering task. A data-driven state feedback control approach was designed through the learning of dynamics model; in such a control scheme, the optimal time-scaling function (i.e., one that makes the vehicle reach the target as quickly as possible) is recovered or approximated through a set of sampled trajectories of unknown vehicle model under the physical constraints imposed by the vehicle. The simulated vehicle model is represented by a damped double integrator whose solution to the TOCP is known if no uncertainties exist (see Appendix A) and the numerical solution for double integrator with given boundary states and bounded acceleration is solved in e.g. [23] as the trajectory optimization via barrier function, if the system is completely known. We exploited and implemented PILCO [16], a data-efficient MBRL, in a simulation and then in a real-world experiment involving a toy car. PILCO has had high performance in benchmark tasks involving low-dimensional state spaces, such as in the control of an inverted pendulum and in cart-pole swing-up; specifically, it has demonstrated unprecedented performance in modeling uncertain system dynamics and optimizing a control policy accordingly. The paper [30] contains an easy introduction of PILCO, and some of its extensions and modifications. As summarized in Algorithm 1, PILCO employs the nonparametric GP for the learning of the unknown dynamics and corresponding uncertainty estimates in a probabilistic dynamics model. PILCO finds the optimal policy parameters which minimizes the expected episodic trajectory cost based on learned probabilistic model. The core elements of the PILCO framework, including dynamics modeling, trajectory prediction, policy evaluation and policy optimization, are briefly described in this section.

Algorithm 1: PILCO [16]

1: *Define* parametrized policy: $\pi : x_t \times \theta \rightarrow u_t$
2: *Initialize* policy parameters $\theta \in N(0, I)$ randomly
3: *Execute* actual system and record initial data
4: **repeat**
5: *Learn* system dynamics model with GP
6: **repeat**
7: *Predict* system trajectories
8: *Evaluate* policy: $J^{\pi}(\theta) = \sum_{t=0}^{T} \gamma^t E_x[\text{cost}(x_t|\theta)]$
9: *Policy improvement*: Update parameter θ by analytic gradient $\frac{dJ^{\pi}(\theta)}{d\theta}$
until convergence θ^* to obtain $\pi^* = \pi(\theta^*)$
10: *Execute* π^* on actual system and record data
11: **until** task completed

3.1 Dynamics modeling

In the real world, model uncertainties and model errors are inevitable in the process of modeling a dynamic system. Various methods have been formulated for modeling and learning unknown system dynamics [13, 25, 26]. In a data-driven setting, system dynamics (instead of representations by differential or difference equations) or vector fields that contain uncertainties, nonlinearities, and disturbances are represented in the form of a set of trajectories obtained from the iterated performace of a mission. Therefore, the model learning algorithm must be able to cope with the uncertainty and noise in the collected trajectory data. PILCO adopts GP probabilistic modeling and inference to learn the transition dynamics of a real-world agent), as represented by a prediction model of the true system dynamics by a probability distribution on a space of transition functions for planning (computing the desired state and control trajectories) and control learning. Therefore, PILCO effectively handles uncertainties and reduces the effect of model errors or simplification on the represented system dynamics, as derived through nontrivial mathematical and physical equations. In this respect, PILCO has eliminated the common drawback of model-based frameworks to some extent. Consider an unknown system described by

$$x_{t+1} = f(x_t, u_t) \text{ with } x_t \in R^D, u_t \in R^F \tag{1}$$

In PILCO, a GP can be used to model the unknown transition function (1). The training inputs are data in the form of state–action (x, u) pairs generated by the unknown transition function

$$\tilde{x}_t = \begin{bmatrix} x_t \\ u_t \end{bmatrix} \in R^{D+F} \tag{2}$$

where $u_t = \pi(x_t, \theta)$ with θ as policy parameters depends on a policy $\pi : R^D \rightarrow R^F$ mapping the perceived state to an action. The training target for model learning is chosen as the delta state (the difference between consecutive states) for predicting the difference between current state and next state given action:

$$\Delta_t = x_{t+1} - x_t \in R^D \tag{3}$$

In this paper, the mean and variance of the prior multivariate Gaussian distribution for f modeled as a GP are chosen to be a zero mean function and squared exponential covariance kernel function, respectively,

$$k(\tilde{x}_i, \tilde{x}_j) = \sigma_f^2 \exp\left(-\frac{1}{2}(\tilde{x}_i, \tilde{x}_j)^T \Lambda^{-1}(\tilde{x}_i, \tilde{x}_j)\right) \tag{4}$$

where the variance of the function σ_f^2 and Λ:=diag$([l_1^2, l_2^2, \dots, l_{D+F}^2])$ depending on the length scales are the hyperparameters. With n training samples $\tilde{X}$:=$[\tilde{x}_1, \dots, \tilde{x}_n]$ and y:=$[\Delta_1, \dots, \Delta_n]$, the posterior GP hyperparameters are learned through evidence maximization and describes a one-step prediction model of x_{t+1} for state trajectory generation from Δ_t and x_t as follows.

posterior state distribution

$$p(x_{t+1}|x_t, u_t) = N(x_{t+1}|\mu_{t+1}, \Sigma_{t+1}) \tag{5}$$

mean

$$\mu_{t+1} = x_t + E_f[\Delta_t] \tag{6}$$

variance

$$\Sigma_{t+1} = \text{Var}_f(\Delta_t) \tag{7}$$

where $E_f[\Delta_t]$ and $\text{Var}_f(\Delta_t)$ can be calculated from prior distribution. In practice, computationally tractable means and variances of GP are used as the most likely estimate for the training data and the confidence in the prediction, respectively, for further decision-making.

3.2 Deterministic trajectory prediction

For the subsequent step of policy evaluation, PILCO first predicts long term system trajectories with the learned transition dynamics, given a policy. The distribution of state x_t at time t is assumed to be Gaussian with mean μ_t and covariance Σ_t where $p(x_t) \sim N(\mu_t, \Sigma_t)$. In order to predict next state x_{t+1}, the distribution $p(\tilde{x}_t)$ and $p(u_t)$ are needed. This is done by assuming $p(u_t)$ is Gaussian and by approximating state-control distribution $p(\tilde{x}_t)$ by a Gaussian with correct mean and variance. The mean and covariance of the predictive control distribution $p(u_t)$ is obtained by integrating out the state from $u_t = \pi(x_t, \theta)$.

$$p(u_t) = \int p(\tilde{x}_t) dx_t \tag{8}$$

The distribution of the change in state Δ_t

$$p(\Delta_t) = \iint p(f(\tilde{x}_t)|\tilde{x}_t) p(\tilde{x}_t) df d\tilde{x}_t \tag{9}$$

is subsequently approximated by a Gaussian distribution with mean μ_{Σ} and variance Σ_{Δ} through calculating of the posterior mean and covariance of $p(\Delta_t)$ by moment matching or linearization [16]. The posterior state distribution in (5–7) can then be approximated by $p(x_{t+1}) \sim N(\mu_{t+1}, \Sigma_{t+1})$ with

$$\mu_{t+1} = \mu_t + \mu_{\Delta} \tag{10}$$

$$\Sigma_{t+1} = \Sigma_t + \Sigma_{\Delta} + \mathrm{Cov}[x_t, \Delta_{t+1}] + \mathrm{Cov}[\Delta_{t+1}, x_t] \tag{11}$$

$$\mathrm{Cov}[x_t, \Delta_{t+1}] = \mathrm{Cov}[x_t, u_t]\Sigma_u^{-1}\mathrm{Cov}[u_t, \Delta_{t+1}] \tag{12}$$

3.3 Policy evaluation (reward function)

The policy is evaluated with the expected return. To this end, the predictive trajectories $\{p(x_t), t = 0, 1, \ldots, N\}$ are retrieved, given a policy, to compute the expected cumulative reward (13, 14) for policy evaluation.

$$J(\theta) = \sum_{t=0}^{T} \gamma^t E_x[\mathrm{cost}(x_t)|\theta] \tag{13}$$

$$E_x[\mathrm{cost}(x_t)|\theta] = \int \mathrm{cost}(x_t) N(x_t|\mu_t, \Sigma_t) dx_t \tag{14}$$

where γ is the future discount factor which determines the importance of future costs on the reward and quantifies the time after which the costs have less influence on the rewards. Note that a large γ will cause the accumulated cost (13) calculated at the end of episode to reduce more slowly in late time as time index $t \to N$. A small γ means current reward is more important than the future rewards, thus would be good for uniform convergence. Using the currently learned model for policy evaluation is the key to data-efficiency of PILCO.

***Remark**. Since the learning reward (13, 14) does not include the control regularization term (such as using L^1, L^2, or mixed L^1-L^2 norm regularization as additional sparsity-inducing cost for (13, 14) [9]), it allows gradient computation for model-based optimization described in the following subsection.

3.4 Cost function design

In general, there are quite a number of cost functions possible for the reward of RL acting as the control for uncertain system, yet the effectiveness of a specific RL algorithm depends on the applications indeed. The objective function can be multimodal to allow different skills to be learned. For example, a parametric cost function [31] can be used to switch the cost from one that follows a quadratic cost function to a time-optimal cost during learning to generate different feedback control schemes in response to different events. Kabzan et al. [32] used a progress-maximizing cost function, defined as one that ensured that the learning agent drove as far as possible within every time step. For TOCP we consider, an appropriate per-step cost at $x_t = x(t)$ at sampling time t is represented in (15)

$$\mathrm{cost}(\mathrm{x}(t)) = 1 - \exp\left(-\frac{1}{2\sigma_c^2}\|\mathrm{x}(t) - \mathrm{x}_{\mathrm{target}}\|_2^2\right) \in [0, 1] \tag{15}$$

where $\mathbf{x}_{\text{target}}$ is the target state and the cost width can be tuned with σ_c. It measures how fast the vehicle progresses on the track to reach the target (subject to tolerance, i.e. the neighborhood of the target state) in terms of exponential function of pairwise Euclidean distance within the episode horizon. The data-driven control (18, 19) for (15) is to maximize the accumulated trajectory cost (13, 14). The expected cost depending on the state, thus on the control parameter θ through the mean and covariance of $u_t = \pi(x_t, \theta)$, allows for analytic integration [16].

3.5 Policy optimization

With the model uncertainty handled by the GP, PILCO employs model-based policy search for planning and uses analytic gradient of cost function for optimization of policy parameter. The policy is improved episodically through the gradient information of (13): the policy update step is in the gradient directions toward high reward region of action space to search for optimal policy (best action sequence) directly. Since the cumulative reward function and transition function are differentiable with respect to the policy parameter θ, analytic gradient $J^\pi(\theta)/d\theta$ with respect to the policy parameter, which depends on the policy parametrization, is available for many interesting control parametrizations, which involves several applications of chain rule [16]. Finally, an advantage of PILCO is that through the analytic expression of the cost function with respect to policy parameter, any standard gradient-based optimization method can be implemented to search directly in the policy space for the optimal policy parameter θ of high (say, thousands) dimension, which minimizes the total cost $J^\pi(\theta)$ so as to obtain desired state trajectory with higher reward.

4. Simulation results and discussions

4.1 Simulation scenario and settings

For the simulation purpose, we consider a simple aggressive driving task whose action space is one-dimensional acceleration in a given interval, and the state space is a two-dimensional space of position and velocity with the position confined on a linear track to constrain the exploration for sample efficiency. The driving scenario is visualized in **Figure 1**, in which the autonomous car is represented by a black box. We used a double integrator with an unknown but constant point mass to simulate the behavior of a real vehicle traveling along a straight line on flat ground with unknown but constant viscous friction. Let $x(t)$ denote the distance traveled by a point mass m on a frictional ground with viscous friction coefficient c controlled by an applied control subject to symmetric constraint $u \in U_{ad} = [-u_{\max}, u_{\max}]$, where the action space U_{ad} is an admissible control set (a convex polytope) that overcomes static friction and avoid slipping. Defining $\mathbf{x} = [x, \dot{x}]^T$ as the state vector of simulated vehicle, the state equations for the vehicle dynamics can be written as

$$\dot{\mathrm{x}} = \mathrm{A}\mathrm{x} + \mathrm{b}u, \mathrm{x}(0) = 0 \tag{16}$$

where

$$A = \begin{bmatrix} 0 & 1 \\ 0 & -\frac{c}{m} \end{bmatrix}, \ b = \begin{bmatrix} 0 \\ \frac{1}{m} \end{bmatrix} \tag{17}$$

Since the vehicle (15) is controllable, the existence of a time-optimal control input with at most one switching to steer the vehicle from a rest state $\mathbf{x}(0) = 0$ at time zero to a neighborhood of another state $\mathbf{x}(t)$ in a given time t is ensured by controllability is ensured by PMP. In Appendix A, we provide the analytical solution of state-to-state transfer TOCP to (15) with explicit parameter dependence. The insights offered by the analytical solution for learning and its performance is that given the symmetric control bound, the control trajectory and switch time are determined by $\frac{c}{m}$ and c (the natural frequency and damping ratio), while the state (position and velocity) trajectory depends on $\frac{c}{m}$, m explicitly and c implicitly via the control.

Settings. For concreteness, the learning scenario is illustrated in **Figure 1**, in which the vehicle is represented by a black box. The vehicle had a horizontal length of 30 cm and a mass of 0.5 kg. In addition, to mimic a real vehicle on the road, we set a symmetric bound of ± 4 m/s^2 on the acceleration control and a friction coefficient of 0.1. The motion began from rest at the origin and proceeded along a linear track for a fixed duration T_{terminal} per an acceleration policy. It is desired that the target state $\mathbf{x}(T) = \mathbf{x}_{\text{target}} = [5, 0]^T$ for a prescribed distance $L = 5$ with a $T \leq T_{\text{terminal}}$ value that is as small as possible (i.e., the policy determines the maximum $\dot{x}$ for each point x on the pre-specified path and a travel time T for task completion), where T_{terminal} is the duration of learning. Crucially, the agent itself has no prior knowledge on the mass or friction coefficient. The simulation is conducted on an Intel Core i7-8700k and 16 GB RAM using MATLAB.

To facilitate the episodic policy search in PILCO, we set each episode to be $T_{\text{terminal}} = 4\ s$, and each episode was further discretized into 40 time steps (i.e., $\Delta t = 0.1\ s$ for each time step). We ran 16 episodes for the agent to learn the task, and the first task was randomly initialized. The code is available at https://github.com/brianhc liao/PILCO_AlphaBot. Our choice of learning time $T_{\text{terminal}} = 4$ (for one episode) in the simulations was a trade-off between computational cost (which increases if more

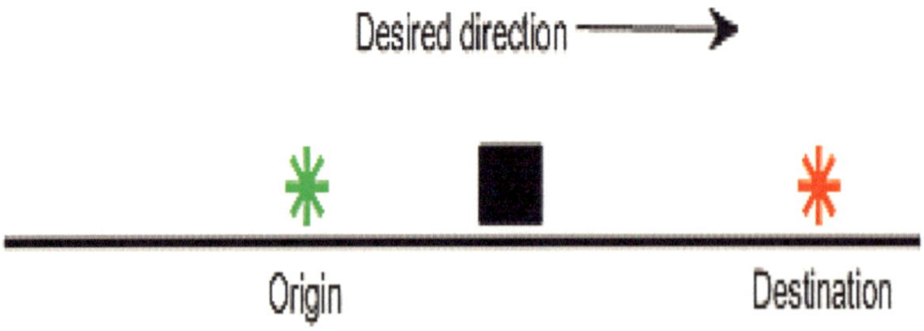

Figure 1.
Setup of one-dimensional state-to-state transfer task. The black box depicts the car, which is modeled using a point-mass double integrator. The car begins moving from the origin (green star) at rest along a straight line to reach the target (red star) along a rough plane. The task involved the execution of different acceleration control policies on the double integrator with embedded uncertainties from the same rest state to reach a target state. The resulting state–input pair and cost at each sampling time point were recorded.

data are collected) and performance, where overly small and large scales of T_{terminal} result in aggressive and relaxed learning, respectively. When each episode was run, the agent considered an episode, in which the vehicle was returned to the same initial state (rest state at $t = 0$) after a policy in the parametric form of (18, 19) to the simulated damped double integrator model (15). In this task, both the state and input could be observed for the data to be collected. In one control cycle and for each maneuver policy, we collected a batch of trajectory data generated by (15) (the trajectory did not violate the acceleration limits) at the sampling times points of $t = 0.1, 0.2, \ldots, 3.9, 4.0$ (in seconds) over the time horizon $[0, 4]$. Therefore, a total of 16 episodes (including an initial random-policy round) were executed, each of which had a total of $N = 40$ samples of state–control pairs. Subsequently, a constant-size data set D^i, $(i = 1, 2, \ldots, 16)$ was constructed, and each of which is composed of a sequence of precisely 40 state–action pairs, where the corresponding state was visited by the vehicle and corresponding control input

$$D^i = \left(\left(\mathrm{x}_1^{(i)}, u_1^{(i)}\right), \ldots, \left(\mathrm{x}_{N_{\text{target}}}^{(i)}, u_{N_{\text{target}}}^{(i)}\right), \ldots, \left(\mathrm{x}_{40}^{(j)}, u_{40}^{(j)}\right)\right) \tag{18}$$

$$u_1^{(i)} = \pi\left(\mathrm{x}_1^{(i-1)}, \theta^{(i-1)}\right), \ldots, u_{40}^{(i)} = \pi\left(\mathrm{x}_{40}^{(i-1)}, \theta^{(i-1)}\right) \tag{19}$$

where $N_{\text{target}} \leq N = 40$ denote the first sampling time at which the car passes through the target region. These 40 state-control pairs collected at different sampling times are correlated via state transition which maps current state and acceleration to the next state. The time-discrete system as the state transition function is obtained by using Euler method to (15) and gets the controllability matrix

$$\mathrm{x}_{n+1}^{(i)} = \mathrm{x}_n^{(i)} + \Delta t \mathrm{A} \mathrm{x}_n^{(i)} + \Delta t \mathrm{b} u_n^{(i)} = [\mathrm{I} + \Delta t \mathrm{A}]\mathrm{x}_n^{(i)} + \Delta t \mathrm{b} u_t^{(i)} \tag{20}$$

$$\mathrm{C} = [\mathrm{I} + \Delta t \mathrm{A}] = \begin{bmatrix} 1 & \Delta t \\ 0 & 1 - \Delta t \dfrac{c}{m} \end{bmatrix} = \text{controllability matrix} \tag{21}$$

where $n = 0, 1, 2, \ldots, 39$, $\left(\mathrm{x}_0^{(i)} = \mathrm{x}_0, u_0^{(i)}\right)$ given, $u_0^{(i)}$ is an initial input for arbitrary initial exploration. Since **C** is nonsingular for any m and c, the system considered is ensemble controllable to guarantee the existence of an appropriate input for steering task, even the system parameters are unknown. The data set D^i in each trial recorded how the vehicle modified its input at each sampled point on the path in accordance with the stage cost at each sampling time point, which was the only type of feedback information that the vehicle received during the vehicle–environment interaction during learning. The i-th episode learning data were to be used for a demonstration in which the reward in the next $(i + 1)$th learning cycle was used to revise the policy for an optimization over action sequences in an effort to minimize the distance between the predicted future state and the target by large progress; this revision was conducted through an online estimation of the vehicle model based on the batch of collected data (for the whole completed episode) during the learning process.

4.2 Data-driven approximate time-optimal control design

We posit the basis functions $\{\phi_i, i = 1, 2, \ldots n_b\}$ where n_b is the total number of basis functions that can be used to represent the policy as a linear combination with a set of parameters over the basis functions [16]; this linear combination is represented as follows:

$$u(\mathrm{x},\theta) = \sum_{t=0}^{n_b} w_i \phi_i(\mathrm{x}) \tag{22}$$

We choose $\phi_i(x)$ in the form of Gaussian Radial Basis Function (RBF)

$$\phi_i(\mathrm{x}) = \exp\left(-\frac{1}{2}(\mathrm{x}-\mu_i)^T \Delta^{-1}(\mathrm{x}-\mu_i)\right) \tag{23}$$

where $\{\mu_i, i = 1, 2, \dots, n_b\}$ are support points, and Δ is covariance matrix. Then $\theta_i = [w_i, \mu_i, \Delta]$ represents the policy parameter vector of the weight, mean, and covariance of each Gaussian RBF. The choice of parametric controller (18, 19) in the form of linear combination of Gaussian RBFs lies in function approximation property and good generalization property, where the controller parameters and the number of basis functions can be optimized using various methods. This implies robustness to vehicle parametric uncertainties and disturbances in learning. Therefore, the class of learning-based parametrized control policy for effectively reject external disturbance and compensate parametric uncertainties we consider is defined with a mapping that maps weights w_i and basis functions ϕ_i to a full (predicted) state (position and velocity) feedback control. Since the optimal control is generally unknown, n_b is chosen to be sufficiently large (100 in simulation) to allow an accurate approximation in model prediction and input-constrained control for the specific state-to-state transfer task.

4.3 Model learning performance and efficiency

PILCO uses GP that relates the policy learning performance to the double integrator with embedded uncertainties [specifically mass m and friction coefficient c in (15)] through its interactions with the environment for model learning. GP maps the policy parameter to the reward, which correlates the policy learning and one-step predictive model trajectory learning. How accurate the learned model trajectories can thus be measured by the reduction in travel time or increase in rewards over episodes, since an accurate dynamic model is required to derive the optimality conditions. In our case, the maneuver time minimization is not directly involved in the transformed cost (13, 14, 19). In fact, via maximizing the expected sum of discounted rewards of progress (or elapsed distance measured along the path) per step by applying admissible acceleration at each time step over the entire episode horizon $[0, T_{\text{terminal}}]$ of learning, the maneuver or control policy tends to achieve a maximum progress per sampling time. Since the decrease of distance to target is directly related to reduction of travel time, it is worth noting that the model learning performance is gradually improved over episode as validated in the cost v.s. the number of episodes (time complexity) plot of **Figure 2** with reduced cost and consistently with the reduced motion duration. This is because decaying factor to the power of t, γ^t in the time-accumulated trajectory cost sum (13) favors the immediate distance cost (19) by maximizing the distance traveled in the first few samples using the discounted factor in the cost function. This promotes the goal-reaching to be achieved with large progression per step at the beginning of the trajectory subject to

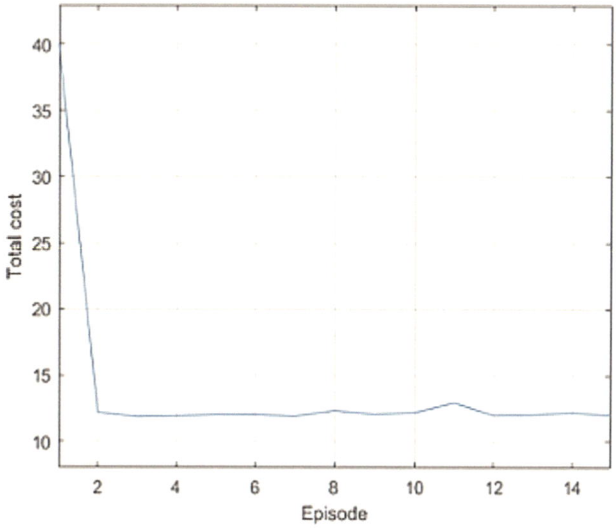

Figure 2.
Graph of total cost against episode. Cost was reduced until (near) convergence was achieved at the second episode and was stable after convergence.

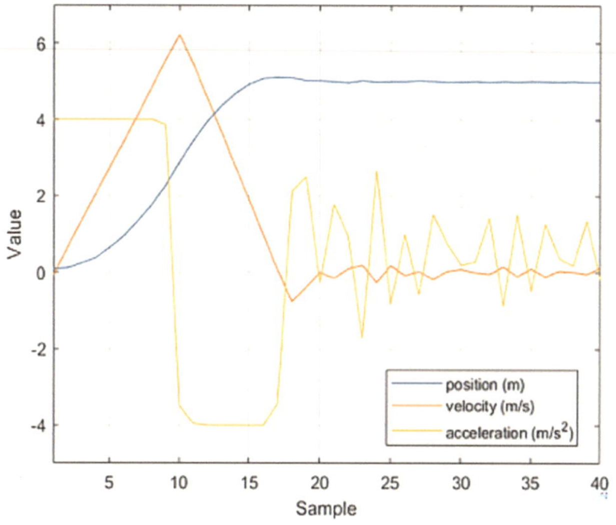

Figure 3.
Converged learning outcome of state (position, velocity) and input (acceleration) trajectories of the vehicle. The velocity profile was triangular and the acceleration input exhibited bang–bang control characteristics. After reaching the neighborhood of the target, the vehicle attempted to stop and remain at the target. A little steady-state oscillation around the target was present because the vehicle could not decelerate fast enough to come to rest at the target. That is, the required fast motion induces overshooting the target if it does not brake in time and then reversed as hard as possible to be closer to the target.

symmetric acceleration bound of the vehicle, thus faster trajectory so as to fit trajectories with higher rewards.

We can see two highly different trajectory behaviors in **Figure 3a**, while **Figure 3b** shows the learning curves in the position-velocity phase plane. The initial motion direction is either aligned with the desired direction toward the target or counterdirectional with it. The first trajectory generated in the first episode was counterintuitive in that the vehicle drove away from the destination. This unusual behavior was associated with a high trajectory cost due to divergence from the target, causing the rest of the trajectories in the second to last episodes to have their directions of motion switched when the vehicle was initially traveling toward the target. We observe that in the second episode, a nearly correct state response is generated to fit the highest reward. The learning behavior from second episode until the final one is relatively identical with small steady-state oscillation due to no terminal cost to control the terminal state at the end of horizon. The task-specific cost (19) is approaching zero as the final state is only around the target (i.e. $\mathbf{x}(T) \to \mathbf{x}_{\text{target}}$), considering the inaccuracy in reaching the target and residual vibration was present after the vehicle reached the neighborhood of the target. The required fast motion induces overshooting the target and when overshoot occurs, turn-back to be closer to the target is observed during learning. The policy is updated and exploration is terminated to maintain the reward at its highest throughout the remaining learning episodes.

4.4 Policy learning performance-verifying time-optimality

Theoretically, in ideal situations of no external disturbances and no model errors, by taking into account the input constraints, the analytical solution can be obtained (see Appendix A) for the damped double integrator when the time-optimal acceleration input is a bang-bang control [20] (i.e., a piecewise constant $\pm u_{\max}$ at all time points that is on the boundary of the admissible control set U_{ad}) with one sign change (see also the following interpretation in Section 5.2). Accordingly, the effectiveness of the learned policy with respect to the time-optimal motion task is measured by analytical solution. As shown in **Figure 4b** and **Table 1**, the learned velocity trajectory is visually converges to a profile which is very similar the analytical solution. We see that the characteristics of exact solution are learned: the learned velocity profile almost coincides with the time-optimal zigzag profile with exactly one switching point whose height and corresponding time are, respectively, the maximum allowed velocity and t_{sw}, and optimal acceleration is applied for each possible state on the linear track under symmetric acceleration constraint. For different c/m and c parameters, the fastest goal-reaching behavior from the same initial rest state results in different zigzag velocity profiles with different travel times for given L and fixed $u_{\max}$. **Table 1**

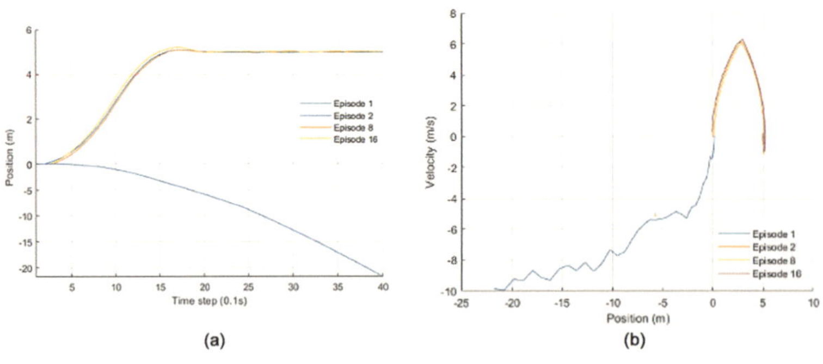

Figure 4.
Learned response in time and phase plane. Episode 1 features the car reversing. In episode 2, a correct, nearly time-optimal motion toward the target was produced. (a) Vehicle position at each episode over 40 sampling time steps spanning a time horizon of 4 s. (b) Predicted position–velocity trajectories in the phase plane through application of data-driven control on learned model for the goal-reaching task along a linear track. The state trajectories became more accurate (time-optimal) with respect to the predicted velocity trajectory as the model was updated during learning. The time-optimal velocity trajectory upon convergence with one instance of switching is shown.

m[a]	c[b]	t_{sw}[c] (s)	T^*[d] (s)
0.5	0.0	0.79057	1.58114
0.5	0.1	0.854717	1.58443
0.5	0.096	0.852088	1.58418
0.5	0.104	0.857352	1.5847
0.48	0.1	0.836147	1.55229
0.52	0.1	0.872974	1.61595

[a] *m is the mass, a given parameter.*
[b] *c is the coefficient of friction, also a given parameter.*
[c] *t_{sw} is the switching time under bang-bang control.*
[d] *T is the minimum time spent in the whole motion.*

Table 1.
This shows t_{sw} and T^* of the triangle profile in the presence of ∓4% deviation of both mass and friction coefficient simultaneously with accurate $c = 0.1$, $m = 0.5$ for given $L = 5$ and $u_{max} = 4$. For comparison, $t_{sw} = 0.79057$ and $T^* = 2t_{sw} = 1.58114$ for the case of $c = 0$. It indicates the friction clearly slows down the fastest motion.

shows the effect of parameter variations due to approximate dynamics model on achievable minimum time. Note that $T^* \neq 2t_{sw}$ when $c \neq 0$. It is because the vehicle should not only have the highest acceleration to reach a highest speed at t_{sw} and decelerate sufficiently fast to meet the null boundary conditions, but also, affected by the viscous frictional force which resists the vehicle's movement speed to increase. The acceleration policy that produces this unique zigzag position-velocity profile corresponding to the bang-bang control with appropriate switching time to meet the boundary conditions is the goal of the policy learning based on learned model in an attempt to match.

In conclusion, the success of the simulation of goal-reaching trajectories along the track (task relevant states in reachable set) rendered the time-optimal trajectories on a straight road feasible. As the simulation shows, the objective function decreases the approximation errors introduced by the currently learned GP model and reduces consistently the travel time very effectively for optimal policy search to fit trajectories with higher rewards, and the resulting policy encodes the desirable spatial and temporal correlations approximately. However, the GP learning results from the collected data set are only valid for the given path and task, but the laws governing physical dynamics apply across paths and tasks. Furthermore, there is discrepancy between the desired time-optimality and the objective of cumulative maximum progress per step we implemented, we observed that the algorithm, despite the natural exploitation–exploration characteristics of the saturating cost function, was sometimes not guaranteed to be globally optimal; it could get stuck in a local optimum because the optimization problem was not convex [16, 18], but still yields good results owing to learning convergence of the generalizable controller as approximate solution to TOCP, whose sensitivity with respect to parameter variations of m and c is small.

5. Sim2Real policy transfer experiment

In real-world experiments, a relatively general uncertain nonlinear vehicle model is either not readily available or overly complicated for the design of state-to-state steering maneuvers. Therefore, to solve this problem, one approach to sim2real

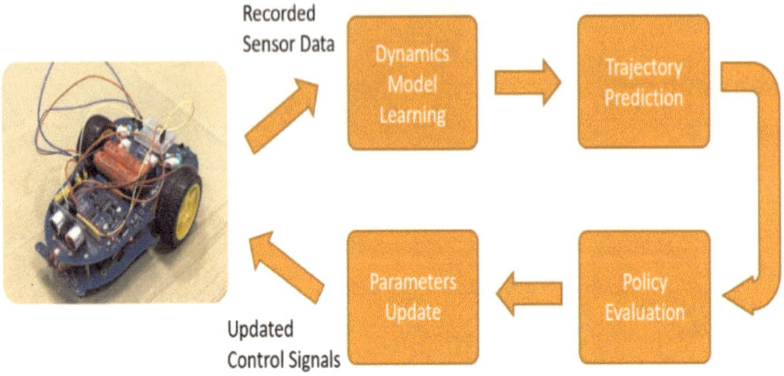

Figure 5.
Low-cost model car AlphaBot for the experiment (left photo) and the experimental setup involved in driving AlphaBot along a linear track.

control is the transfer of skills learned from a simple simulated model to a similar real system executing a similar task. To test the learned policy in a real-world environment and as detailed in this section, we conducted a sim2real validation experiment with a low-cost Raspberry Pi–controlled small car, called AlphaBot, as shown in **Figure 5**. The simple car was equipped with photo-interrupters and ultrasonic sensors for state measurements. It could be controlled to either accelerate or brake. Because the robot unit was inexpensive, the low pulse-width-modulation duty cycles that were translated from the control signals might be incapable of driving the car, which also makes the task more complex. Before we conducted the experiment, we verified that the robot had sufficient power to travel a distance of L along the linear track to reach the target. In contrast to the simulation scenario, the experimental setup had an additional velocity limit for the car, which functioned as a state constraint. The velocity limit stemmed from the voltage constraints of the board and was therefore not directly handled by the control signal.

5.1 Setup

A small car was designated to travel from a point 180 cm from the wall to a point 50 cm from the wall. Its heading was fixed at a forward-facing angle of 0° for straight line motion. The task characteristics (system dynamics along a linear track on a frictional plane) and the environment for learning in the simulation and experiment were very similar. We assumed that the model of the car was similar to that in the experiment; it was a second-order dynamics model with slightly changed parameters. This model is universal because it is based on fundamental physical principles. Thus, we could reuse the same RBF controller (18, 19) that was used in the simulated double integrator (15) as a time-optimal feedback control of the actual vehicle. When applied to a real vehicle, the learned policy from simulation runs thus can be viewed as an optimal demonstration by the simulated system (15). This policy provides a good understanding of the time-optimal motion of the vehicle motion with only an input constraint under no disturbance and no state constraint. In the experiment, the control signal generated from the RL algorithm ranged from -2 to 2 and was translated into the rate of change of the duty cycles of the motor PWM signal on board.

DOI: http://dx.doi.org/10.5772/intechopen.103690

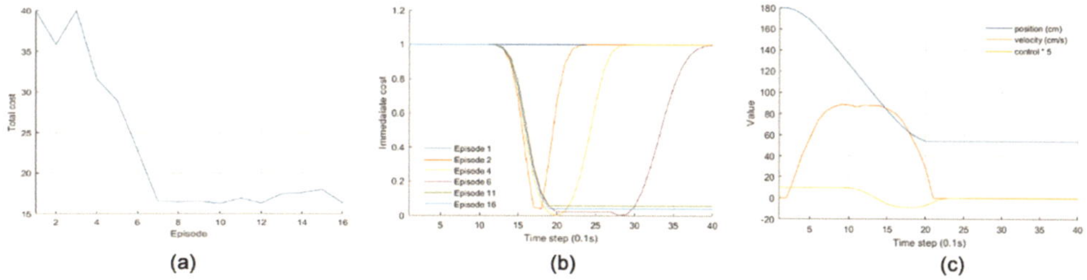

Figure 6.
Experimental results. (a) Graph of total cost against episodes. Trajectory cost decreased after some transients until convergence at the seventh episode where a travel time of approximately 2 s was achieved. (b) Intermediate costs in selected episodes. (c) Outcome with respect to learning state and control trajectories.

5.2 Results

As indicated in **Figure 6**, the sim2real transfer experiment, which applied the same data-driven control learned from simulation, worked well on the low-cost car. The vehicle drove at each point on the path at its highest allowable velocity; this observation was similar to that in the simulation. This validates the slight generalization capability (robustness and stability) of the MBRL with Gaussian RBF kernel under similar dynamics and task constraints. The learning curve is illustrated in **Figure 6a**. The total cost was 40 at the initial trial of the task for the saturating cost function. The vehicle attained a lowest cost of approximately 16 and a travel time of approximately 2.2 s at the seventh episode with a total experience time of 28 s. This approach was deemed to be very efficient because learning was successful after only a small number of trials. In **Figure 6b**, immediate cost at every time step (per-step cost) at various episodes is plotted, showing the learning process and indicating that the arrival time decreased over episodes. The decrease was, however, not monotonic; thus, the intermediate trajectories prior to the completion of learning may not be acceptable until a nearly time-optimal control input is finally obtained at convergence, which was the desired control goal. As indicated in **Figure 6c**, by following a nearly time-optimal velocity along the track, the final position of the car, despite not overshooting the target or fluctuating, was slightly off the target by 50 cm partly because of modeling uncertainties and localization errors from lateral tracking error and sensor inaccuracies. A velocity limit was set because of the electronic voltage constraints on the robot and was therefore not directly handled by the robot control signal.

5.3 Discrepancies between simulation and experiment

Prior knowledge about a detailed description of the vehicle dynamics that contain uncertainties is not available or is not required in both simulation and sim2real experiment. Instead, the model is learnt through the Gaussian Processes to reduce the accumulated cost, thus contributing to decrease the maneuver time. However, the convergence in the simulation tends to require less learning episodes to converge, which possibly stems from the following major differences. The first important difference is that the learning experiment is on the basis of more complicated vehicle dynamics. In fact, the inherent factors confronted during real world experiment that cannot be neglected or accounted for in a simple physical model (18). These can affect the control performance include uncertainties caused by lateral drift and wheel sideslip of the vehicle, physical properties such as inertia and complicated, hard to model friction characteristics (such as Coulomb friction proportional to sign($\dot{x}$) and aerodynamic (quadratic) drag force proportional to $\dot{x}^2$, in addition to viscous friction proportional to $\dot{x}$), or unknown external forces inherent in the vehicle-terrain interaction as a result of the inaccuracy of sensor measurements, motor characteristics and torque disturbances and variation of environment interactions such as uneven ground. These factors make the prediction of the sampled states on the linear track under the input ambiguous and affect the stability of system to reach the target. Considerations of vehicle hardware also entail an additional velocity limit (due to a limit to how much voltage the hardware can take) that ensures that rapid motion causes no harm or instability on the vehicle at each trial. These factors result in differing state trajectories in the experiment and simulation under the same control. Due to low sensitivity of TOCP with respect to system parameters, however, the convergence of learning to the approximate solution is ensured under these factors. Therefore, a new approximate optimal solution, if the system deviated from the initially demonstrated trajectory, is recovered by learning controller.

5.4 Interpretations of learning results

In a scenario of short-distance driving on a linear path, we see that data-driven control (18, 19) realizing the nearly time-optimal state-to-state steer control in a simulated model is demonstrated as a good approximated time-optimal control of actual system with similar dynamics and task characteristics. The error which is in sim2real experiment between the learned simulated model from the dataset and learned near time-optimal control performance can be split into two aspects. The first one is the error between the actual state $\mathrm{x}_a(t)$ and the theoretical time-optimal state $\mathrm{x}_{\mathrm{opt}}(t)$. The second one is the error between the theoretical time-optimal state $\mathrm{x}_{\mathrm{opt}}(t)$ and the learned simulated state x_t. Let $\left(\mathrm{x}_{\mathrm{opt}}(t), u_{\mathrm{opt}}(t) = u\left(\mathrm{x}_{\mathrm{opt}}(t), \theta_{\mathrm{opt}}(t)\right)\right)$ be the time-optimal state-input pair of simulation model (15). For a successful sim2real experiment, we assume that the simulation model (15) can be extended to the actual system (24) in real experiment

$$\dot{\mathrm{x}}_a = \mathrm{A}\mathrm{x}_a + \mathrm{b}u_{\mathrm{opt}} + \delta\mathrm{A}\mathrm{x}_a \tag{24}$$
$$\mathrm{x}_a(0) = 0, u \in U_{ad} = [-u_{\max}, u_{\max}]$$

where $\mathrm{x}_a(t)$ is the actual state, and

$$\delta\mathrm{A} = \begin{bmatrix} 0 & 0 \\ 0 & \dfrac{\delta c}{m} \end{bmatrix} \tag{25}$$

is a bounded additive disturbance due to the error δc of actual frictions and simulated friction settings with other a priori unknown factors that linearly related to x_a, $u_{\max}$ is the maximum control limitation (or state constraint) confined by hardware setting. Note (24) applies the same time-optimal control (18) learned from the simulated system. We assume the error δc is bounded by $k_c > 0$. Since x_{opt} satisfies (15), we then have

$$\dot{\mathrm{x}}_a - \dot{\mathrm{x}}_{opt} = \mathrm{A}\left(\mathrm{x}_a - \mathrm{x}_{opt}\right) + \delta\mathrm{A}\mathrm{x}_a \tag{26}$$

Integrating (26) and moving associated terms, we have

$$x_a - x_{opt} = e^{At}\int_0^t e^{-As}(\delta A x_a)ds \tag{27}$$

for $0 < t < T_{\text{terminal}}$. Since T_{terminal} is finite, x_a is bounded in $[0, T_{\text{terminal}}]$ (note, (24) can be solved analytically), the difference $|x_a - x_{\text{opt}}|$ is proportional to k_c with a constant C, independent of t. Therefore, the error between the actual state $x_a(t)$ and the theoretical time-optimal state $x_{\text{opt}}(t)$ is arbitrarily small if k_c tends to 0.

On the other hand, the dynamics system with transition function is discretized by Euler method as

$$x_{t+1} = x_t + \Delta t A x_t + \Delta t b u_t \tag{28}$$

Since this is the dynamic system learned by PILCO, thus the difference between x_t and x_{opt} is proportional to $(\Delta t)^2$ with constant C, independent of t. Therefore, as $k_c, \Delta t \to 0$, we have $x_a \to x_{\text{opt}}$.

6. Conclusions

A case study of TOCP, whose solution typically requires a known model with given boundary conditions for applying PMP, is framed as a MBRL task for rest-to-rest steering along a linear path by a vehicle that was modeled as an uncertain double integrator subject to constant acceleration bound. An important aspect of this paper is a novel application of PILCO, an existing data-efficient MBRL, to approximately solve TOCP for uncertain damped double integrator without accurate parameters. Feasibility of learning convergence is empirically verified by parametric sensitivity of exact time-optimal solution. The consistency of learned velocity profile closer to that obtained by analytical time-optimal control is shown by simulation first and then implemented on a sim2real experiment. The learned velocity can be further revised to account for a velocity limit through scaling without the need of replanning or relearning for performing less aggressive maneuvers. Our case study expands the scope of problems that can be successfully solved by MBRL (specifically, PILCO), serving as a robust adaptive optimal control, without prior parametric model representation, and it demonstrates the capability in compensating for uncertainties and external disturbances, which can cause the state trajectories to deviate from the optimal simulated state trajectory. For the challenging problem of learning a safe velocity for various road topologies and traffic flows, MBRL suffers from the accumulated compounding error over long horizon. And the comparison with other learning approaches to solution of optimal control problems is our future work.

Acknowledgements

This work was supported by an internal funding from Institute of Information Science, Academia Sinica, Taipei, Taiwan.

Appendix A

Instead of data-driven assessment by comparing with state of the art algorithms, an analytical time-optimal solution is provided to validate the learned trajectory solution. The system of the TOCP to damped double integrator can be rewritten as

$$\frac{d^2x}{dt^2} = -\frac{c}{m}\frac{dx}{dt} + \frac{1}{m}u(t) \tag{29}$$

where

$$\begin{aligned} x(0) = 0, x(T) = L \\ \dot{x}(0) = 0, \dot{x}(T) = 0 \\ |u(t)| \leq u_{\max} \end{aligned} \tag{30}$$

Integrating the Eq. (29), we then get

$$\dot{x}(t) - \dot{x}(0) = -\frac{c}{m}(x(t) - x(0)) + \frac{1}{m}(U(t) - U(0)) \tag{31}$$

where

$$U(t) = U(0) + \int_0^t u(s)ds \tag{32}$$

Reducing the Eq. (31) by invoking the boundary conditions, we get

$$\dot{x}(t) = -\frac{c}{m}x(t) + \frac{1}{m}(U(t) - U(0)) \tag{33}$$

Thus,

$$x(t) = x(0) + e^{-\frac{c}{m}t}\int_0^t \frac{e^{\frac{c}{m}s}}{m}(U(t) - U(0))ds \tag{34}$$

The state trajectory with fixed initial state can be expressed by the control $u(t)$, thereby removing the system dynamics constraint.

From the terminal condition at T, we get

$$L = e^{-\frac{c}{m}T}\int_0^T \frac{e^{\frac{c}{m}s}}{m}(U(t) - U(0))ds \tag{35}$$

Therefore,

$$\begin{aligned} Le^{\frac{c}{m}T} &= \int_0^T \frac{e^{\frac{c}{m}s}}{m}\left(\int_0^s u(\tilde{s})d\tilde{s}\right)ds = \int_0^T\int_0^s \frac{e^{\frac{c}{m}s}}{m}u(\tilde{s})d\tilde{s}ds \\ &= \int_0^T\int_s^T \frac{e^{\frac{c}{m}s}}{m}u(\tilde{s})dsd\tilde{s} = \int_0^T\left(u(\tilde{s})\int_s^T \frac{e^{\frac{c}{m}s}}{m}ds\right)d\tilde{s} = \frac{1}{c}\int_0^T u(\tilde{s})\left(e^{\frac{c}{m}T} - e^{\frac{c}{m}s}\right)d\tilde{s} \\ &= \frac{1}{c}e^{\frac{c}{m}T}\int_0^T u(s)ds - \frac{1}{c}\int_0^T u(s)e^{\frac{c}{m}s}ds = \frac{1}{c}e^{\frac{c}{m}T}(U(T) - U(0)) - \frac{1}{c}\int_0^T u(s)e^{\frac{c}{m}s}ds \end{aligned} \tag{36}$$

And substituting into (33) with boundary condition of $\dot{x}(T) = 0$, we get

$$U(T) = Lc + U(0) \tag{37}$$

Therefore,

$$\int_0^T u(s)e^{\frac{c}{m}s}ds = 0 \tag{38}$$

Thus, the conditions (32), (37) and (38) can be equivalently converted into

$$\begin{cases} \int_0^T u(s)ds = Lc \\ \int_0^T u(s)e^{\frac{c}{m}s}ds = 0 \end{cases} \tag{39}$$

(39) together with another condition $|u(t)| \le u_{\max}$ on $[0, T]$, these three are complete constraints on $u(t)$ of the problem. Now, according to bang-bang control, the solution $u(t)$ should be $\pm u_{\max}$ for T to achieve the minimum. Since (29) is a linear second-order system, the bang-bang control has exactly one switching time. Let t_{sw} denote the switching time and T^* the minimum time possible. Based on this pattern for $u(t)$,

$$u(t) = \begin{cases} u_{\max} & \text{if } t \in [0, t_{sw}] \\ -u_{\max} & \text{if } t \in [t_{sw}, T^*] \end{cases} \tag{40}$$

we can define

$$u(t) = c_1 I_{[0,t_{sw}]} + c_2 I_{[t_{sw},T^*]} \tag{41}$$

where

$$I_{[a,b]}(x) = \begin{cases} 1 & \text{if } x \in [a,b] \\ 0 & \text{otherwise} \end{cases} \tag{42}$$

Substituting (41) into (39), it becomes

$$\begin{cases} c_1 t_{sw} + c_2(T - t_{sw}) = Lc \\ c_1 \frac{m}{c}\left(e^{\frac{c}{m}t_{sw}}\right) + c_2 \frac{m}{c}\left(e^{\frac{c}{m}T} - e^{\frac{c}{m}t_{sw}}\right) = 0 \end{cases} \tag{43}$$

Therefore, the c_1 and c_2 are

$$\begin{cases} c_1 = \dfrac{Lc\left(e^{\frac{c}{m}T} - e^{\frac{c}{m}t_{sw}}\right)}{t_{sw}\left(e^{\frac{c}{m}T} - 1\right) - T\left(e^{\frac{c}{m}t_{sw}} - 1\right)} \\ c_2 = \dfrac{-Lc\left(e^{\frac{c}{m}t_{sw}} - 1\right)}{t_{sw}\left(e^{\frac{c}{m}T} - 1\right) - T\left(e^{\frac{c}{m}t_{sw}} - 1\right)} \end{cases} \tag{44}$$

According to $|u(t)| \leq u_{max}$ and bang-bang principle, we have

$$\begin{cases} c_1 = \dfrac{Lc\left(e^{\frac{c}{m}T} - e^{\frac{c}{m}t_{sw}}\right)}{t_{sw}\left(e^{\frac{c}{m}T} - 1\right) - T\left(e^{\frac{c}{m}t_{sw}} - 1\right)} = u_{max} \\ c_2 = \dfrac{-Lc\left(e^{\frac{c}{m}t_{sw}} - 1\right)}{t_{sw}\left(e^{\frac{c}{m}T} - 1\right) - T\left(e^{\frac{c}{m}t_{sw}} - 1\right)} = -u_{max} \end{cases} \tag{45}$$

Thus, $\left(e^{\frac{c}{m}T} - e^{\frac{c}{m}t_{sw}}\right) = \left(e^{\frac{c}{m}t_{sw}} - 1\right)$ and we get $t_{sw} = \frac{m}{c} \ln\left(\frac{e^{\frac{c}{m}T}+1}{2}\right)$. Now, substituting it into

$$c_1 = \frac{Lc\left(e^{\frac{c}{m}T} - e^{\frac{c}{m}t_{sw}}\right)}{t_{sw}\left(e^{\frac{c}{m}T} - 1\right) - T\left(e^{\frac{c}{m}t_{sw}} - 1\right)} = u_{max} \tag{46}$$

we get the minimum case of T, i.e. $T = T^*$. (This can be computed by numerical methods. See results in **Table 1**).

Author details

Hsuan-Cheng Liao†, Han-Jung Chou† and Jing-Sin Liu*†
Institute of Information Science, Taipei, Taiwan, ROC

*Address all correspondence to: liu@iis.sinica.edu.tw

† These authors contributed equally.

References

[1] Ostafew CJ, Schoellig AP, Barfoot TD, Collier J. Speed daemon: Experience-based mobile robot speed scheduling. In: Canadian Conference on Computer and Robot Vision. USA: IEEE; 2014. pp. 56-62

[2] Rao AV. Trajectory optimization: A survey. In: Optimization and Optimal Control in Automotive Systems. Cham: Springer; 2014. pp. 3-21

[3] Bobrow J, Dubowsky S, Gibson J. Time-optimal control of robotic manipulators along specified paths. International Journal of Robotics Research. 1985;**4**(3):3-17

[4] Verscheure D, Demeulenaere B, Swevers J, DeSchutter J, Diehl M. Time-optimal path tracking for robots: A convex optimization approach. IEEE Transcation on Automatic Control. 2009;**54**(10):2318-2327

[5] Tohid A, Norrlöf M, Löfberg J, Hansson A. Convex optimization approach for time-optimal path tracking of robots with speed dependent constraints. IFAC Proceedings Volumes. 2011, 2011;**44**(1):14648-14653

[6] Shin K, McKay N. Selection of near-minimum time geometric paths for robotic manipulators. IEEE Transactions on Automatic Control. 1986;**31**(6): 501-511

[7] Wigstrom O, Lennartson B, Vergnano A, Breitholtz C. High-level scheduling of energy optimal trajectories. IEEE Transactions on Automation Science and Engineering. 2013;**10**(1):57-64

[8] Bianco CGL, Romano M. Optimal velocity planning for autonomous vehicles considering curvature constraints. In: IEEE International Conference on Robotics and Automation. USA: IEEE; 2007. pp. 2706-2711

[9] Dinev T, Merkt W, Ivan V, Havoutis I, Vijayakumar S. Sparsity-inducing Optimal Control Via Differential Dynamic Programming. USA: IEEE; 2020. arXiv preprint arXiv: 2011.07325

[10] Kunz T, Stilman M. Time-optimal trajectory generation for path following with bounded acceleration and velocity. In: Proceedings of Robotics Science and Systems VIII. Cambridge, Massachusetts, United States: MIT Press; 2012. pp. 1-8

[11] Jond HB, Nabiyev VV, Akbarimajd A. Planning of mobile robots under limited velocity and acceleration. In: 22nd Signal Processing and Communications Applications Conference. USA: IEEE; 2014. pp. 1579-1582

[12] Pham Q. A general, fast, and robust implementation of the time-optimal path parameterization algorithm. IEEE Transactions on Robotics. 2014;**30**(6): 1533-1540

[13] Polydoros AS, Nalpantidis L. Survey of model-based reinforcement learning: Applications on robotics. Journal of Intelligent & Robotic Systems. 2017; **86**(2):153-173

[14] Kober J, Bagnell JA, Peters J. Reinforcement learning in robotics: A survey. The International Journal of Robotics Research. 2013;**32**(11): 1238-1274

[15] Dulac-Arnold G, Levine N, Mankowitz DJ, Li J, Paduraru C,

Gowal S, et al. Challenges of real-world reinforcement learning: Definitions, benchmarks and analysis. Machine Learning. 2021;**110**(9):1-50

[16] Deisenroth M, Rasmussen CE. PILCO: A model-based and data-efficient approach to policy search. In: 28th International Conference on Machine Learning (ICML-11). Bellevue, WA, USA: ICML; 2011. pp. 465-472

[17] Martinez-Marin, T. (2005). Learning optimal motion planning for car-like vehicles. IEEE International Conference on Computational Intelligence for Modelling, Control and Automation IEEE USA pp.601-612

[18] Saha O, Dasgupta P, Woosley B. Real-time robot path planning from simple to complex obstacle patterns via transfer learning of options. Autonomous Robots. 2019:1-23

[19] Hartman G, Shiller Z, Azaria A. Deep reinforcement learning for time optimal velocity control using prior knowledge. In: IEEE 31st International Conference on Tools with Artificial Intelligence (ICTAI). USA: IEEE; 2018. arXiv preprint arXiv:1811.11615

[20] Liberzon D. Calculus of Variations and Optimal Control Theory: A Concise Introduction. Princeton University Press; 2011

[21] Ozatay E, Ozguner U, Filev D. Velocity profile optimization of on road vehicles: Pontryagin's maximum principle based approach. Control Engineering Practice. 2017;**61**:244-254

[22] Stryk O, Bulirsch R. Direct and indirect methods for trajectory optimization. Annals of Operation Research. 1992;**37**(1):357-373

[23] Hauser J, Saccon A. A barrier function method for the optimization of trajectory functionals with constraints. In: Proceedings of the 45th IEEE Conference on Decision and Control. USA: IEEE; 2006. pp. 864-869

[24] Qian X, Navarro I, de La Fortelle A, Moutarde F. Motion planning for urban autonomous driving using Bézier curves and MPC. In: IEEE 19th International Conference on Intelligent Transportation Systems (ITSC). USA: IEEE; 2016. pp. 826-833

[25] Song C, Boularias A. Identifying Mechanical Models Through Differentiable Simulations. Ithaca, New York: Cornell University; 2020. arXiv preprint arXiv:2005.05410

[26] Geist AR, Trimpe S. Structured learning of rigid-body dynamics: A survey and unified view. GAMM-Mitteilungen. 2020;**44**(2):e202100009. arXiv preprint arXiv:2012.06250

[27] Moerland TM, Broekens J, Jonker CM. Model-based Reinforcement Learning: A Survey. Ithaca, New York: Cornell University; 2020. arXiv preprint arXiv:2006.16712

[28] Liu M, Chowdhary G, Da Silva BC, Liu SY, How JP. Gaussian processes for learning and control: A tutorial with examples. IEEE Control Systems Magazine. 2018;**38**(5):53-86

[29] Pineda L, Amos B, Zhang A, Lambert NO, Calandra R. MBRL-LIB: A Modular Library for Model-based Reinforcement Learning. Ithaca, New York: Cornell University; 2021. arXiv preprint arXiv:2104.10159. Available from: https://github.com/facebookresearch/mbrl-lib

[30] Brunzema P. Review on Data-Efficient Learning for Physical Systems

using Gaussian Processes. Berlin, Germany: ResearchGate; 2021. Available from: researchgate.net

[31] Sprague CI, Izzo D, Ögren P. Learning a Family of Optimal State Feedback Controllers. Ithaca, New York: Cornell University; 2019. arXiv preprint arXiv:1902.10139

[32] Kabzan J, Hewing L, Liniger A, Zeilinger MN. Learning-based model predictive control for autonomous racing. IEEE Robotics and Automation Letters. 2019;**4**(4):3363-3370

Chapter 6

Self-Supervised Contrastive Representation Learning in Computer Vision

Yalin Bastanlar and Semih Orhan

Abstract

Although its origins date a few decades back, contrastive learning has recently gained popularity due to its achievements in self-supervised learning, especially in computer vision. Supervised learning usually requires a decent amount of labeled data, which is not easy to obtain for many applications. With self-supervised learning, we can use inexpensive unlabeled data and achieve a training on a pretext task. Such a training helps us to learn powerful representations. In most cases, for a downstream task, self-supervised training is fine-tuned with the available amount of labeled data. In this study, we review common pretext and downstream tasks in computer vision and we present the latest self-supervised contrastive learning techniques, which are implemented as Siamese neural networks. Lastly, we present a case study where self-supervised contrastive learning was applied to learn representations of semantic masks of images. Performance was evaluated on an image retrieval task and results reveal that, in accordance with the findings in the literature, fine-tuning the self-supervised training showed the best performance.

Keywords: self-supervised learning, contrastive learning, representation learning, computer vision, deep learning, pattern recognition

1. Introduction

For an effective training, supervised learning requires a decent amount of labeled data, which is expensive. Unlabeled and inexpensive data (e.g. text and images on the Internet) is considerably more than the limited size datasets labeled by humans. We can use unlabeled data and perform a training on a pretext task, which is a **self-supervised** approach since we do not use the labels in our real task. Although the task and the defined loss are not the ones in our actual objective, we can still learn some representations that are valuable enough to be used for the final task. We basically learn a parametric mapping from the input data to a feature vector or tensor. In most cases, a smaller amount of labeled data is used to fine-tune the self-supervised training.

Although its origins date as back as 1990s [1, 2], contrastive learning has recently gained popularity due to its achievements in self-supervised learning, especially in computer vision. In **contrastive learning**, a representation is learned by comparing among the input samples. The comparison can be based on the similarity between positive pairs or dissimilarity of negative pairs. The goal is to learn such an embedding space in which similar samples stay close to each other while dissimilar ones are far apart. Contrastive learning can be applied to both supervised and unsupervised settings. Let us consider image classification problem. In supervised setting, positive pairs are different instances with the same label and negative samples are selected from other labels (**Figure 1**). On the other hand, in unsupervised (or self-supervised) setting, positive pairs are parts (or augmented versions) of the same instance and negative samples are other instances with any label. Khosla *et al.* [3] provide a performance comparison between supervised and self-supervised training for image classification problem. Also, a more comprehensive review of contrastive learning can be found in [4].

Since a self-supervised model does not know the actual labels corresponding to the inputs, its success depends on the design of the pretext tasks to generate the pseudo-labels from part of the input data. With these pseudo-labels, training on pretext task is performed with a 'supervised' loss function. Final performance on the pretext task is not important, but we hope that the learned intermediate representations can capture good information and be beneficial to a variety of downstream tasks.

Especially in computer vision and natural language processing (NLP), deep learning has become the most popular machine learning approach [5]. In parallel, self-supervised learning studies in computer vision have employed CNNs. **Figure 2** shows the knowledge transfer from a self-supervised training to a supervised one in a deep learning setting. We save convolutional layers which are assumed to produce learned representations. We change/add fully connected layers, place a classifier head and train with the limited amount of labeled data for a downstream task like image classification or object detection.

The remainder of this chapter is structured as follows. Pretext tasks that are common in literature are reviewed in Section 2. Section 3 has detailed information about recent self-supervised learning models that use Siamese architectures. Section 4 provides our own experimental study where self-supervised contrastive learning is employed to learn representations of semantic segmentation masks, which is followed by the conclusions in Section 5.

Figure 1.
Self-supervised (left) vs. supervised (right) contrastive learning. Training results in an embedding space such that similar sample pairs stay close to each other while dissimilar ones are far apart. Figure is reproduced based on [3].

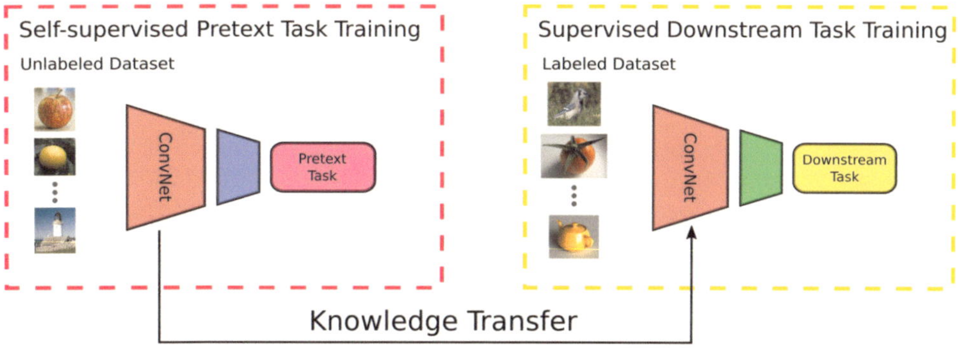

Figure 2.
A model is first trained with a pretext task with unlabeled data, then fine-tuned on the downstream task with limited amount of labeled data. Usually convolution layers, which are mostly responsible of learning representations, are transferred. A few fully-connected layers towards the end are changed or retrained.

Figure 3.
Several random transformations applied to a patch from the unlabeled dataset to be used for self-supervised learning. Original sample is in top-left. The idea was first used by [6] and the figure is from the original paper with author's permission.

2. Pretext tasks for self-supervised learning

Image Distortion. We expect that when an image goes through a small amount of distortion, its semantic meaning does not change. Dosovitskiy *et al.* [6] used this idea to create a exemplar-based classification task, where a surrogate class is formed with each dataset sample by applying a variety of transformations, namely translation, scaling, rotation, contrast and color (**Figure 3**). When this approach is applied to whole image instances, it can be called as 'instance discrimination' [7], where augmented versions of the same image (positive pair) should have similar representations and augmented versions of the different images (negative pair) should have different representations.

This is not only one of the first pretext tasks but also a very popular one. We will see in Section 3 that the mentioned type of augmentations have succeeded in learning useful representations and have achieved state-of-the-art results in transfer learning for downstream computer vision tasks.

Image Rotation. Each input image is first rotated by a multiple of 90° at random. A model is trained to predict the amount of rotation applied [8]. In basic setting, it is a 4-class classification problem, but different versions can be conceived. To estimate the amount of rotation, this pretext task forces the model learn semantic parts of objects,

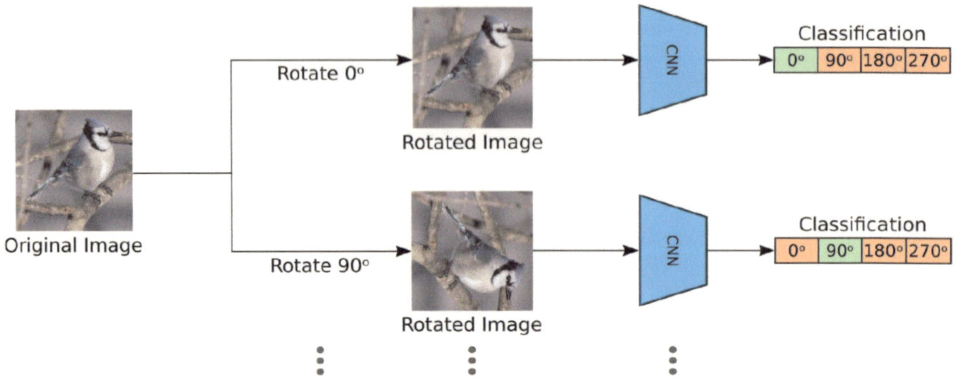

Figure 4.
Self-supervised representation learning by rotating input images, implemented in [8]. The model classifies the rotation [0°, 90°, 180°, 270°].

such as arms, legs, eyes. Thus, it would serve well for a downstream task like object recognition (**Figure 4**).

Jig-saw Puzzle. Noroozi and Favaro [9] designed a jigsaw puzzle game, where a CNN model is trained to place 9 shuffled patches back to the original locations. Each patch is processed independently with shared weights and a probability vector estimated per patch. Then, these estimations were merged to output a permutation.

Image Colorization. The task is to colorize gray-scale images into colorful images [10]. A CNN is trained to predict the colorized version of the input (**Figure 5**). Obtaining a training dataset is inexpensive since training pairs can be easily generated. Model's latent variables represent grayscale images and can be useful for a variety of downstream tasks.

Image Inpainting. The pretext task is filling in a missing piece in the image (e.g. Pathak *et al*. [11]). The model is trained with a combination of the reconstruction (L2) loss and the adversarial loss. It has an encoder-decoder architecture and encoder part can be considered as representation learning.

Last two pretext tasks (image colorization and inpainting) and some other GANs (e.g. image super-resolution [12]) are generation-based methods, where a missing info in the content is generated from available input. Whereas distortion, rotation and jigsaw are context-based self-supervision methods. For more detailed literature on pretext tasks, we refer the readers to the review in [13].

In our study, we concentrate on the context-based approach. Taking advantage of contrastive learning, this approach nowadays achieves state-of-the-art performance

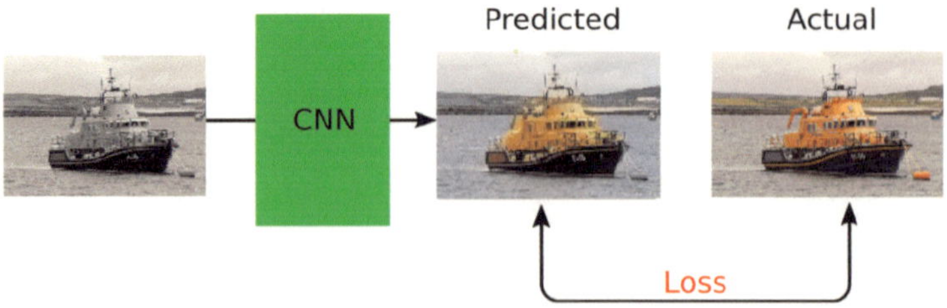

Figure 5.
A model is trained to predict the colorized version of grayscale images (obtaining the dataset is inexpensive).

[14–17]. We will go into details, especially the models with Siamese architecture, in Section 3. The generation-based and context-based method distinction also exists for video representation learning. In [18], an encoder network is used to learn video representations. Then, a decoder uses the representations to predict future frames. Differently, Qian *et al.* [19] employ contrastive learning with distortions (augmentations) to learn representations and to classify video clips.

The rest of our chapter will consider works on image data. Before proceeding, let us give a few examples where contrastive learning is used for image-text pairs. Contrastive Language-Image Pre-training (CLIP, [20]) is a pretext task, where a text encoder and an image encoder are jointly trained to match captions with images. Training set consists of 400 million (image,text) pairs and an inter-modal contrastive loss is defined such that image and text embeddings of same objects will be closer to each other. Then, this pretraining is employed for a downstream task of zero-shot class prediction from images. Li *et al.* [21] performed a similar task for semantic segmentation. An image encoder is trained with a contrastive objective to match pixel semantic embeddings to the text embeddings. Another example presents contrastive learning of medical visual representations from paired images and text [22].

3. Self-supervised contrastive learning models

The goal of contrastive learning is to learn such an embedding space in which similar sample pairs stay close to each other while dissimilar ones are far apart. Implemented using Siamese networks, recent approaches create two different augmentations of samples and feed into the networks for contrastive learning. While SimCLR [14] and MoCo [15] use the negative samples directly along with the positive ones, BYOL [16] and SimSiam [17] achieved similar performance just with the positive samples. Differently, SwAV [23] forced consistency between cluster assignments of augmentations, instead of comparing features directly. Shortly after, vision transformers were included in self-supervised learning architectures [24, 25]. According to the results, not only image classification, but also object detection and semantic segmentation as downstream tasks benefit from self-supervised contrastive learning. Let us briefly explain some of these main approaches.

3.1 SimCLR

Let us describe SimCLR [14] first, then we will describe other methods by comparing to previous ones. SimCLR uses both positive and negative samples, but being positive or negative does not correspond to actual class labels. Augmented versions of the anchor are taken as positives, whereas samples belong to different instances are taken as negatives (**Figure 6**).

- Let T be the set of image transformation operations where $t \sim T$ and $t' \sim T$ are two different transformation operators independently sampled from x. These transformations are random cropping and resizing, random Gaussian blur and random color distortion. A $(\tilde{x}_i, \tilde{x}_j)$ pair of query and key views is positive when these two views are created by applying different transformations on the same image x: $\tilde{x}_i = t(x)$ and $\tilde{x}_j = t'(x)$.

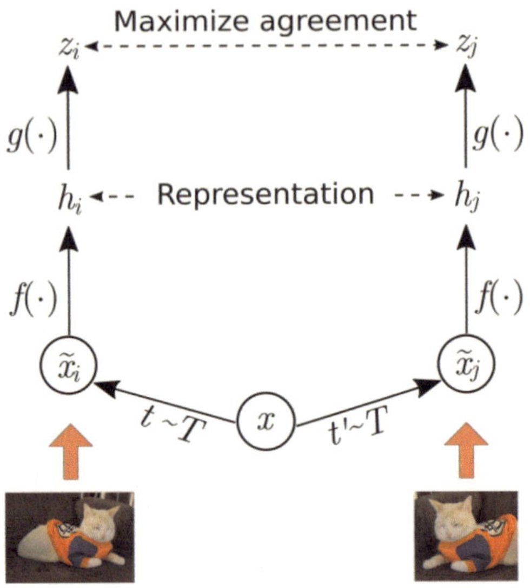

Figure 6.
SimCLR framework [14], where two separate data augmentations are sampled from a predefined family of augmentations (crop, blur color jitter). An encoder network $f(\cdot)$ and a projection head $g(\cdot)$ are trained to maximize the agreement between the embeddings of these two samples. When the self supervised learning is over, projection head can be thrown away.

- A base feature encoder $f(\cdot)$ then extracts the representations from all the augmented data samples $h_i = f(\tilde{x}_i), h_j = f(\tilde{x}_j)$. There is no restriction on the choice of the encoder's architecture but a ResNet-50 [26] model was preferred for SimCLR due to its simplicity. The representation h in this case is the output of the average pooling layer of ResNet-50.

- Each representation h is then fed into a projection head $g(\cdot)$ to map representations to the embedding space where the contrastive loss is applied. $z_i = g(h_i), z_j = g(h_j)$. This projection head can be as simple as a one-layer multi-layer perceptron (MLP) using a non-linear activation.

- A batch of (z_i, z_j) pairs representing the embeddings from two augmented versions of the same image, are then fed into the contrastive loss function which encourages the distance between embeddings from positive pairs to be small and the distances of embeddings from negative pairs to be large.

SimCLR uses the contrastive loss given in Eq. (1). This is a categorical cross-entropy loss to identify the positive sample among a set of negative samples (inspired from InfoNCE [27]).

$$L^{self} = \sum_{i \in I} L_i^{self} = -\sum_{i \in I} \log \frac{\exp(z_i \cdot z_j / \tau)}{\sum_{a \in A(i)} \exp(z_i \cdot z_a / \tau)} \tag{1}$$

N images are randomly taken from the dataset. Thus, the training batch consists of $2N$ images to which data augmentations are randomly applied. Let $i \in I \equiv \{1 \ldots 2N\}$ be the index of an arbitrary augmented sample, then j is the index of the other augmentation of the same original image. $\tau \in R^+$ is a scalar temperature parameter, $\cdot$ represents the dot product, and $A(i) \equiv I - \{i\}$. We call index i the anchor, index j is the positive, and the other $2(N-1)$ indices as negatives. The denominator has a total of $2N-1$ terms (one positive and $2N-2$ negatives).

A common protocol to evaluate self-supervised model efficiency is to place a linear classifier on top of (frozen) layers learnt by self-supervised training and train it for the downstream task with the labeled data. If the performance gap between this self-supervised encoder + linear classifier and a fully-supervised model is small, then the self-supervised training considered as efficient. An alternative evaluation protocol uses semi-supervised learning, i.e. pretrained network is re-trained as a whole with a certain percentage of available labels. Experiments reveal that re-training with only 10% of the labeled data achieves a performance (92.8%) very close to fully-supervised training performance on the whole dataset (94.2%) as reported in [14] (performances are top-5 classification accuracy on ImageNet dataset for ResNet-50).

3.2 Momentum contrast (MoCo)

Contrastive methods based on InfoNCE loss tend to work better with high number of negative examples since negative examples may represent underlying distribution more efficiently. SimCLR requires large batches (4096 samples) to ensure that there is enough negatives which demands high computation power (8 V100 GPUs in their study). To alleviate this need, MoCo [15] uses a dictionary of negative representations that is structured as a FIFO queue. This queue-based dictionary enables us to reuse representations of immediately preceding mini-batches of data. Thus, the main advantage of MoCo compared to SimCLR is that MoCo decouples the batch size from the number of negatives. SimCLR requires a large batch size and suffers performance drops when the batch size is reduced.

Given a query sample x^q, we get a query representation through our online encoder $q = f_q(x^q)$. A list of key representations $\{k_0, k_1, k_2, \ldots\}$ coming from the dictionary and are encoded by a different encoder $k_i = f_k(x_i^k)$ as shown in **Figure** 7. Naming two

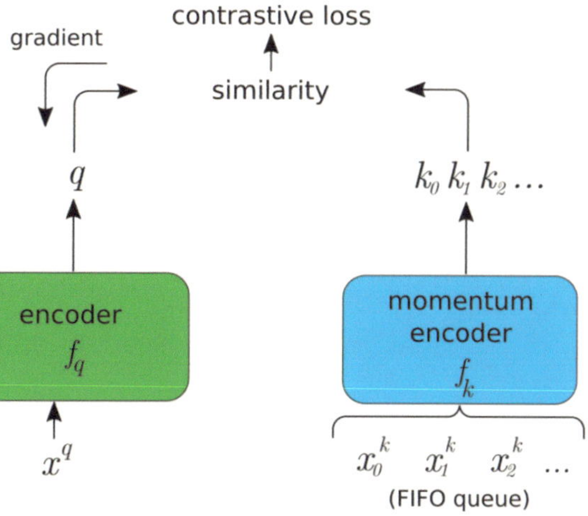

Figure 7.
MoCo framework [15]. The encoder that takes negative samples (from a FIFO queue) is not updated by backpropagation but with the other encoder's parameters with a momentum coefficient. That's why it is called the momentum encoder.

compared representations as query and key is new, but one can think of them as the two augmentations of the same sample in SimCLR (**Figure 6**).

Let us assume that there is a single positive key, k_+, in the dictionary that matches q. Then, the contrastive loss with one positive and $N-1$ negative samples becomes:

$$L_{MoCo} = -\log \frac{\exp\left(q \cdot k_+ / \tau\right)}{\sum_1^N \exp\left(q \cdot k_i / \tau\right)} \tag{2}$$

From the two encoders defined above, for f_k we can not apply backpropagation since it works on the queue. Copying online encoder's (f_q) weights to f_k could be a solution, however MoCo proposed to use a momentum-based update with a momentum coefficient:

$$\theta_k \leftarrow m\theta_k + (1-m)\theta_q \tag{3}$$

where $m \in [0, 1]$ is the momentum coefficient and θ_q and θ_k are parameters of f_q and f_k respectively (**Figure 7**).

Later on, two design choices in SimCLR, namely MLP projection head and more stronger data augmentation were integrated into the approach resulting in MoCo-v2 [28].

3.3 Bootstrap your own latent (BYOL)

Different from the approaches above, BYOL [16] achieves similar representation performance without using negative samples. It relies on two different neural networks (in contrast to SimCLR but similar to MoCo), referred to as online and target networks that interact. Online network has a predictor head. Target network has the same network architecture with the online network except for the predictor head (**Figure 8**). Parameters of the target network are not updated with back-propagation, but with a moving average of online network's weights just as MoCo did for the momentum encoder.

It is curious that how the model escapes from collapsing (i.e. a trivial solution of fixed vector for each sample) when no negative samples are used. Authors of BYOL thought it is due to the momentum update, but later (with SimSiam [17]) it was discovered that using stop-gradient and predictor head is enough.

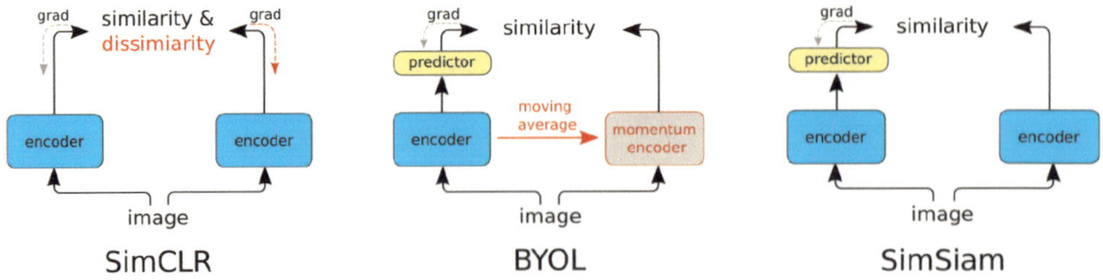

Figure 8.
Comparison of some Siamese architectures: SimCLR, BYOL and SimSiam. Dashed lines indicate back-propagation. Components colored in red are no more needed in SimSiam. Figure is reproduced based on [17].

3.4 Simple Siamese (SimSiam)

BYOL needs to maintain two copies of weights for the two separate networks which can be resource demanding. SimSiam [17] solves this problem with parameter sharing between the networks (with and w/o predictor head). The encoder $f(\cdot)$ shares weights while processing two views. A prediction MLP head, denoted as $g(\cdot)$ transforms the output of only one view. Thus, two augmented views (x_1 and x_2) results in two outputs: $p_1 = g(f(x_1))$ and $z_2 = f(x_2)$. Their negative cosine similarity is denoted as $D(p_1, \text{stopgrad}(z_2))$, where stopgrad operation is an important component. It implements that z_2 is treated as a constant term and encoder receives no gradients from z_2. Gradient only flows back to the encoder through the prediction head.

Finally, negative cosine similarity based total loss is computed in a symmetric fashion:

$$L_{SimSiam} = \frac{1}{2}D(p_1, \text{stopgrad}(z_2)) + \frac{1}{2}D(p_2, \text{stopgrad}(z_1)) \quad (4)$$

Figure 8 compares SimSiam with SimCLR and BYOL. SimSiam [17] does not use negative samples as SimCLR and MoCo did. Success with SimSiam also shows that momentum encoder (or any sort of moving average update of weights) is not needed. Stop-gradient operation and including predictor head are enough to prevent the model from collapsing.

SimSiam also presents transfer learning results for object detection and semantic segmentation downstream tasks. Results reveal that starting with a self-supervised pre-training on ImageNet outperforms image classification pre-training on ImageNet.

3.5 Self-supervised vision transformers

Caron *et al.* [24] proposed another Siamese architecture where one of the network's parameters are updated with a moving average of other's parameters. More interestingly, they replaced encoder CNNs with vision transformers and reported increasing success for various downstream tasks. Shortly after, Li *et al.* [25] proposed a more efficient vision transformer architecture together with a new pre-training task which is based on region matching.

4. Case study: semantic mask representation learning

As a case study, we employ self-supervised contrastive learning to learn representations of semantically segmented images, i.e. semantic masks. This learning task is especially useful when two scenes are compared according to their semantic content. A use case would be image retrieval based localization, where standard approach extract features from RGB images and compare them to find the most similar image in the database [29, 30]. Recently, several studies showed that checking semantic resemblance between query and database images and using this similarity score while retrieving images improves localization accuracy [31–33]. The reason of improvement is that there is appearance difference between images taken at different times (query-database) due to illumination differences, viewpoint variations, seasonal changes.

Figure 9.
The image on top-left was taken in 2008 and the image on top-right was taken in 2019 (source: Google street view) which respectively represent query and database for image retrieval. Observe illumination differences, viewpoint variations and changing objects. Bottom row shows their semantic segmentation results. Semantic similarity can help to verify/deny the localization result.

Although RGB image features are directly affected by those changes, semantic labels are stable most of the time (**Figure 9**).

Given a semantic mask, obtaining the most similar result among the alternatives is not a trivial task. SIFT-like features do not exist to match. Moreover, two masks of the same scene are far from being identical not only because of changing content but also due to camera position and viewpoint variations. Thus, instead of employing a pixel-by-pixel label comparison score, a trainable semantic feature extractor is preferable.

Measuring semantic similarity to distinguish if two images belong to the same scene or not is a task especially suitable for self-supervised learning. Because datasets has to be prepared such that query and database are the same scene but different images (preferably long-term difference) is not easy. However, large amount of semantic masks can easily be obtained for a self-supervised training. We do not need groundtruth masks, since a successful estimation is enough to compute semantic similarity.

4.1 Dataset and self-supervised contrastive training

Our unsupervised learning dataset composed of 3484 images randomly taken from UCF dataset [34]. These are perspective images obtained from Google Street View panoramas which where taken in Pittsburgh, PA before 2014. Our supervised training and test datasets have query-database image pairs. Query images were also taken from UCF dataset (not coinciding with the 3484 images mentioned above). Database images were collected again from Google Street View panoramas at the same locations of query images but in 2019. This time gap results in seasonal changes and illumination variances. Also, a wide camera baseline between the database and query images conforms better to the long-term localization scenario [35]. Top row in **Figure 9** shows an example of query-database image pair with time difference.

Since our aim to learn representations for semantic masks, we first automatically generated a semantic mask for each image in our dataset using a well-performing CNN model [36]. The CNN model we employed trained on Cityscapes [37], which is an urban scene understanding dataset consists of 30 visual classes. Examples are in **Figure 9** (bottom row). After this point, we only have semantic masks in our dataset.

We used SimCLR [14] as our contrastive learning model and trained a ResNet-18 as the encoder. Encoder network (**Figure 10**) produces $h = Enc(x) \in R^{512}$ features, whereas projection network produces $z = Proj(h) \in R^{512}$ features. We set batch size as 85 and resized semantic mask to 64 × 80 resolution (due to GPU memory limitation) and used two different data augmentation methods during the training: random resized crop and random rotation. We set maximum rotation parameter as 3°, since severe rotations are not expected between query and database images. Crop parameter, however, is important to represent the variation in our dataset. Results for varying crop parameter values will be discussed in Section 4.2. Augmentation of semantic masks is visualized in **Figure 10**. Other augmentations (such as color jitter, horizontal flip, brightness and contrast distortions), which are common for image classification and object detection downstream tasks are not included since they are not expected distortions for semantic masks. We used AMSGrad optimizer which is a variant of Adam.

CNN model, trained as explained above, is now ready to produce a similarity score when two semantic masks (one query and one database) are given. After self-supervised training, same network can be fine-tuned with a labeled dataset (query and database segmentation masks for the same scene). For this purpose, we prepared a dataset of 368 query images with their corresponding database images and extracted their semantic masks. **Figure 9** shows an example of this preparation. Not surprisingly, this paired dataset is much smaller than the self-supervised training dataset. Here, common practice in literature is that the projection head (**Figure 10**) is removed

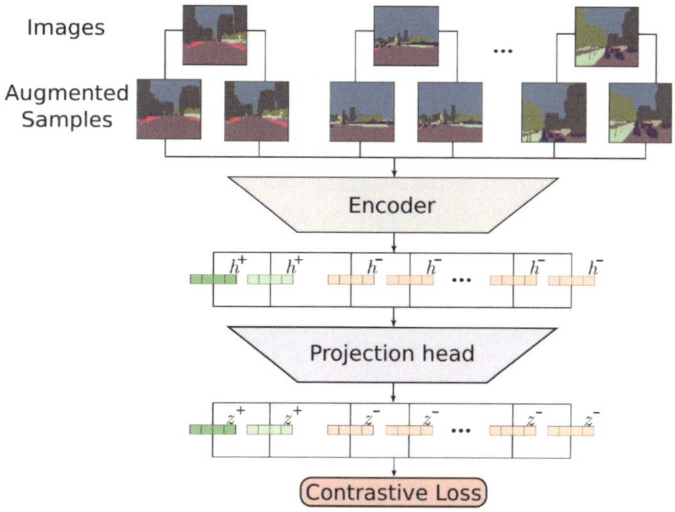

Figure 10.
Illustration of training a CNN model with self-supervised contrastive loss on a dataset that consists of semantically segmented masks. A positive pair is created from two randomly augmented views of the same mask, while negative pairs are created from views of other masks. All masks are encoded by the a shared encoder and projection heads before the representations are evaluated by the contrastive loss function.

after pretraining and a classifier head is added and trained with labeled data for the downstream task. However, our pretext and downstream tasks are the same. We learn semantic representations by treating each sample as its own class (exemplar-CNN [6], instance discrimination [7]). Thus, we do not place a classifier head, but we retrain the network (partially or full).

4.2 Experimental results

To be able to measure the capability of representing semantic masks, we conduct experiments that compare the retrieval accuracies of three training schemes. First is the CNN model which is trained with the supervised training set (368 query-database pairs). This is the baseline model that does not exploit self-supervised training at all. Second is the CNN model that is trained in a self-supervised fashion with 3484 individual semantic masks (no matching pairs). Lastly, the model with self-supervised training is retrained with the supervised training set. Two versions exist: *i*) only replacing dense layers and training them, *ii*) retraining all layers.

Trained models are tested on a test set which consists of 120 query-database pairs (different from 368 pairs used in training). Performances are compared with Recall@N metric. According to this metric, for a query image, the retrieval is considered successful if any of top-N retrieved database images is a correct match. In other words, Recall@1 is the recall when only the top-most retrieval is checked.

We observe in **Table 1** that, only supervised training is not very successful. In fact, for certain N values self-supervised training managed to outperform supervised training alone. This shows the power of self-supervised learning when a large dataset is provided. Our unlabeled dataset is much larger than the labeled dataset ($3484 \gg 368$). Regarding the two fine-tuning schemes, replacing dense layers and training them from scratch improved self-supervised training but not for all N values. On the other hand, fine-tuning all layers worked best by a considerable margin. Since our pretext and downstream tasks are the same (i.e. we do not train a classification head etc.), it is not surprising that replacing dense layers did not help much. **Figure 11** shows several examples where supervised training fails but the proposed self-supervised approach (after fine-tuning) succeeds.

Table 2 presents the effect of minimum crop ratio parameter used in data augmentation module. Since it is an important parameter to represent the variation in our semantic masks, we compare the performance for minimum crop ratio from 0.9 to 0.1. Apart from individual Recall@N values, we also compute and plot mean recall (mean of all N values) in **Table 2** last column and in **Figure 12**. We observe that it is

Training methods	**Retrieval accuracy (Recall@N)**				
	N = 1	**N = 2**	**N = 3**	**N = 4**	**N = 5**
Only supervised training	0.500	0.608	0.767	0.808	0.817
Only self-supervised training	0.567	0.692	0.733	0.775	0.800
Dense layers were replaced and trained	0.542	0.675	0.767	0.825	0.850
All layers were fine-tuned	**0.633**	**0.758**	**0.808**	**0.858**	**0.867**

Table 1.
Only supervised training is compared with self-supervised training and fine-tuned versions of it.

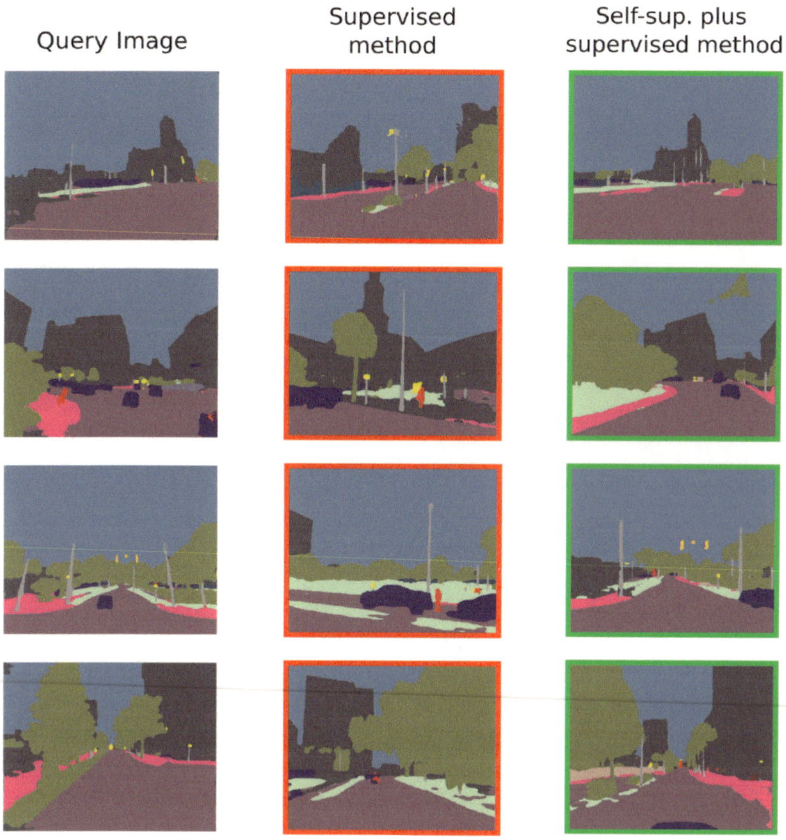

Figure 11.
Each row shows a retrieval result for a given query (left column). Examples show the cases where only supervised training (middle column) fails at Recall@1, but utilizing self-supervised training and then fine-tuning on the labeled dataset (query-database pairs) correctly retrieves (last column).

Crop ratio	Retrieval accuracy (Recall@N)					
	N = 1	N = 2	N = 3	N = 4	N = 5	mean
0.90	0.608	0.708	0.758	0.817	0.858	0.750
0.80	0.617	0.733	0.800	0.848	0.867	0.773
0.70	0.617	0.742	**0.817**	**0.858**	**0.875**	0.782
0.60	**0.633**	**0.758**	0.808	**0.858**	0.867	**0.785**
0.50	0.617	0.700	0.767	0.808	0.833	0.745
0.40	0.575	0.692	0.717	0.783	0.825	0.718
0.30	0.567	0.675	0.742	0.767	0.808	0.712
0.20	0.542	0.633	0.717	0.767	0.783	0.688
0.10	0.525	0.608	0.675	0.742	0.783	0.667

Table 2.
Effect of the minimum crop ratio parameter in data augmentation at the stage of retraining of the self-supervised model.

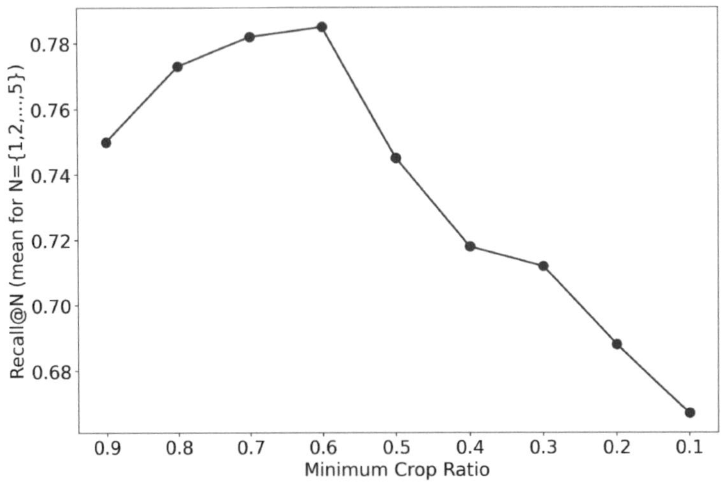

Figure 12.
Mean Recall@N values for varying min. Crop ratio parameter. Observe the reverse U-shape with a peak at 0.6.

highest around 0.6 and 0.7. Performance gradually drops as we increase or decrease the minimum crop ratio. A minimum random crop parameter of 0.6 means that cropped mask covers at least 60% area of the original mask. Since query and database masks in our training and test datasets have a considerable overlap ratio, it is reasonable that 0.6 or higher overlaps serve best. This result is also in accordance with the finding in [38] that there is a reverse U-shape relationship between the performance and the mutual information within augmented views. When crops are close to each other (high mutual information, e.g. crop ratio = 0.9) the model does not benefit from them much. On the other hand, for low crop ratios (low mutual information) model can not learn well since views look quite different from each other. Peak performance stays somewhere in between.

5. Conclusions

In this chapter, we presented the main concepts in self-supervised contrastive learning and reviewed the approaches that attracted attention due to their success in computer vision. Contrastive learning that aims to end up in an embedding space where similar samples stay close to each other was implemented successfully with Siamese neural networks. Necessity on huge computation power was also alleviated with the most recent models. Currently, for common downstream tasks of computer vision such as object detection and semantic segmentation, self-supervised pre-training is a better alternative than using a model trained on ImageNet for image classification.

We also presented a case study where self-supervised contrastive learning is applied to learn representations of semantic masks of images. Performance was evaluated on an image retrieval task where the most similar semantic mask is retrieved from the database for a given query. In compliance with the results on other vision tasks in the literature, fine-tuning the self-supervised model with available labeled data gave better results than the supervised training alone.

Acknowledgements

This work was supported by the Scientific and Technological Research Council of Turkey (TÜBİTAK) under Grant No. 120E500 and also under 2214-A International Researcher Fellowship Programme.

Abbreviations

CNN	Convolutional neural network
RGB	Red green blue
NLP	Natural language processing
LSTM	Long short term memory
GAN	Generative adversarial network
MoCo	Momentum contrast
BYOL	Bootstrap your own latent
MLP	Multi-layer perceptron
FIFO	First in first out

Author details

Yalin Bastanlar* and Semih Orhan
Department of Computer Engineering, Izmir Institute of Technology, Izmir, Turkey

*Address all correspondence to: yalinbastanlar@iyte.edu.tr

References

[1] Bromley J, Guyon I, LeCun Y, Sackinger E, Shah R. Signature verification using a siamese time delay neural network. Processing Systems. Advances in Neural Information Processing Systems. 1993:737-744

[2] Becker S, Hinton GE. Self-organizing neural network that discovers surfaces in random-dot stereograms. Nature. 1992; **355**:161-163

[3] Khosla P, Teterwak P, Wang C, Sarna A, Tian Y, Isola P, et al. Supervised contrastive learning. Advances in Neural Information Processing Systems. 2020

[4] Le-Khac PH, Healy G, Smeaton AF. Contrastive representation learning: A framework and review. IEEE Access. 2020

[5] Tekir S, Bastanlar Y. Deep learning: Exemplar studies in natural language processing and computer vision. In: Data Mining - Methods, Applications and Systems. London: InTechOpen; 2020. DOI: 10.5772/intechopen.91813

[6] Dosovitskiy A, Springenberg JT, Riedmiller M, Brox T. Discriminative unsupervised feature learning with convolutional neural networks. Advances in Neural Information Processing Systems. 2014

[7] Wu Z, Xiong Y, Yu SX, Lin D. Unsupervised feature learning via non-parametric instance discrimination. In: Proceedings of the IEEE Conference on Computer Vision and Pattern Recognition (CVPR). 2018

[8] Gidaris S, Singh P, Komodakis N. Unsupervised representation learning by predicting image rotations. ICLR. 2018

[9] Noroozi M, Favaro P. Unsupervised learning of visual representations by solving jigsaw puzzles. European Conference on Computer Vision (ECCV). 2016

[10] Zhang R, Isola P, Efros A. Colorful Image Colorization. European Conference on Computer Vision (ECCV). 2016

[11] Pathak D, Krahenbuhl P, Donahue J, Darrell T, Efros AA. Context encoders: Feature learning by inpainting. In: Proceedings of the IEEE Conference on Computer Vision and Pattern Recognition (CVPR). 2016

[12] Ledig C, Theis L, Huszar F, Caballero J, Cunningham A, Acosta A, et al. Photo-realistic single image super-resolution using a generative adversarial network. In: Proceedings of the IEEE Conference on Computer Vision and Pattern Recognition (CVPR). 2017

[13] Jing L, Tian Y. Self-supervised visual feature learning with deep neural networks: A survey. IEEE Transactions on Pattern Analysis and Machine Intelligence. 2021;**43**(11): 4037-4058. DOI: 10.1109/TPAMI.2020.2992393

[14] Chen T, Kornblith S, Norouzi M, Hinton G. A simple framework for contrastive learning of visual representations. In: International Conference on Machine Learning (ICML). 2020

[15] He K, Fan H, Wu Y, Xie S, Girshick R. Momentum contrast for unsupervised visual representation learning. In: Proceedings of the IEEE Conference on Computer Vision and Pattern Recognition (CVPR). 2020

[16] Grill JB, Strub F, Altché F, Tallec C, Richemond PH, Buchatskaya E, et al. Bootstrap your own latent: A new approach to self-supervised learning. Advances in Neural Information Processing Systems. 2020

[17] Chen X, He K. Exploring simple siamese representation learning. In: Proceedings of the IEEE/CVF Conference on Computer Vision and Pattern Recognition (CVPR). 2021

[18] Srivastava N, Mansimov E, Salakhutdinov R. Unsupervised learning of video representations using LSTMs. International Conference on Machine Learning. 2015

[19] Qian R, Meng T, Gong B, Yang MH, Wang H, Belongie S, et al. Spatiotemporal contrastive video representation learning. In: Proceedings of the IEEE/CVF Conference on Computer Vision and Pattern Recognition (CVPR). 2021

[20] Radford A, Kim JW, Hallacy C, Ramesh A, Goh G, Agarwal S, et al. Learning transferable visual models from natural language supervision. Proceedings of the 38th International Conference on Machine Learning. 2021

[21] Li B, Weinberger KQ, Belongie S, Koltun V, Ranftl R. Language-driven semantic segmentation. International Conference on Learning Representations (ICLR). 2022

[22] Zhang Y, Jiang H, Miura Y, Manning CD, Langlotz CP. Contrastive learning of medical visual representations from paired images and text. arXiv:2010.00747v1. 2020

[23] Caron M, Misra I, Mairal J, Goyal P, Bojanowski P, Joulin A. Unsupervised learning of visual features by contrasting cluster assignments. Advances in Neural Information Processing Systems (NeurIPS). 2020

[24] Caron M, Touvron H, Misra I, Jegou H, Mairal J, Bojanowski P, et al. Emerging properties in self-supervised vision transformers. International Conference on Computer Vision (ICCV). 2021

[25] Li C, Yang J, Zhang P, Gao M, Xiao B, Dai X, et al. Efficient self-supervised vision transformers for representation learning. International Conference on Learning Representations (ICLR). 2022

[26] He K, Zhang X, Ren S, Sun J. Deep residual learning for image recognition. In: Proceedings of the IEEE Conference on Computer Vision and Pattern Recognition (CVPR). 2016

[27] van den Oord A, Li Y, Vinyals O. Representation learning with contrastive predictive coding. arXiv:1807.03748. 2018

[28] Chen X, Fan H, Girshick R, He K. Improved baselines with momentum contrastive learning. arXiv:2003.04297. 2020

[29] Arandjelovic R, Gronat P, Torii A, Pajdla T, Sivic J. Net VLAD: CNN architecture for weakly supervised place recognition. In: Proceedings of the IEEE Conference on Computer Vision and Pattern Recognition (CVPR). 2016

[30] Ge Y, Wang H, Zhu F, Zhao R, Li H. Self-supervising fine-grained region similarities for large-scale image localization. European Conference on Computer Vision (ECCV). 2020

[31] Cinaroglu I, Bastanlar Y. Long-term image-based vehicle localization improved with learnt semantic descriptors. Engineering Science and

Technology, an International Journal. 2022;**35**:101098

[32] Cinaroglu I, Bastanlar Y. Training semantic descriptors for image-based localization. ECCV Workshop on Perception for Autonomous Driving. 2020

[33] Orhan S, Guerrero JJ, Bastanlar Y. Semantic pose verification for outdoor visual localization with self- supervised contrastive learning. Proceedings of the IEEE Conference on Computer Vision and Pattern Recognition (CVPR) Workshops. 2022

[34] Zamir AR, Shah M. Image geo-localization based on multiple nearest neighbor feature matching using generalized graphs. IEEE Transactions on Pattern Analysis and Machine Intelligence. 2014;**36**(8):1546-1558

[35] Sattler T, Maddern W, Toft C, Torii A, Hammarstrand L, Stenborg E, et al. Benchmarking 6DOF outdoor visual localization in changing conditions. In: Proceedings of the IEEE Conference on Computer Vision and Pattern Recognition (CVPR). 2018

[36] Sun K, Zhao Y, Jiang B, Cheng T, Xiao B, Liu D, et al. High-resolution representations for labeling pixels and regions. Proceedings of the IEEE Conference on Computer Vision and Pattern Recognition (CVPR). 2019

[37] Cordts M, Omran M, Ramos S, Rehfeld T, Enzweiler M, Benenson R, et al. The cityscapes dataset for semantic urban scene understanding. In: Proceedings of the IEEE Conference on Computer Vision and Pattern Recognition (CVPR). 2016

[38] Tian Y, Sun C, Poole B, Krishnan D, Schmid C, Isola P. What makes for good views for contrastive learning? Advances in Neural Information Processing Systems (NeurIPS). 2020

Chapter 7

Analysis of Brain Computer Interface Using Deep and Machine Learning

Nabil Ajali-Hernández and Carlos M. Travieso-Gonzalez

Abstract

Pattern recognition is becoming increasingly important topic in all sectors of society. From the optimization of processes in the industry to the detection and diagnosis of diseases in medicine. Brain-computer interfaces are introduced in this chapter. Systems capable of analyzing brain signal patterns, processing and interpreting them through machine and deep learning algorithms. In this chapter, a hybrid deep/machine learning ensemble system for brain pattern recognition is proposed. It is capable to recognize patterns and translate the decisions to BCI systems. For this, a public database (Physionet) with data of motor tasks and mental tasks is used. The development of this chapter consists of a brief summary of the state of the art, the presentation of the model together with some results and some promising conclusions.

Keywords: brain-computer interfaces, deep learning, machine learning, pattern recognition, artificial intelligence, neural network

1. Introduction

The brain is the most important organ in the human body. It processes, integrates and coordinates the information it receives from the organs and the senses and makes decisions, sending them to the rest of the body, like a processor in a computer. The brain works through electrochemical impulses, called synapses, which allow the transmission of information between neurons [1–3].

These impulses could be classified by their frequency into different types of brain waves. Delta (1–3 Hz), theta (3–8 Hz), alpha (8–13 Hz), beta (13–30 Hz), and gamma (30–100 Hz) waves [4]. These brain waves are the reflection of electrical activity (in microvolts) and therefore of thoughts and motor intentions. They can be captured by the electroencephalogram (EEG) and their study can lead to the detection of pathologies related to the brain (Alzheimer's, Parkinson's, epilepsy) [5, 6].

Figure 1 shows the EEG of a normal person, where 64 channels have been placed throughout the head (10–20 system) and brain waves are monitored over time. It can be seen how there are an infinity of patterns that provide important information about what is happening.

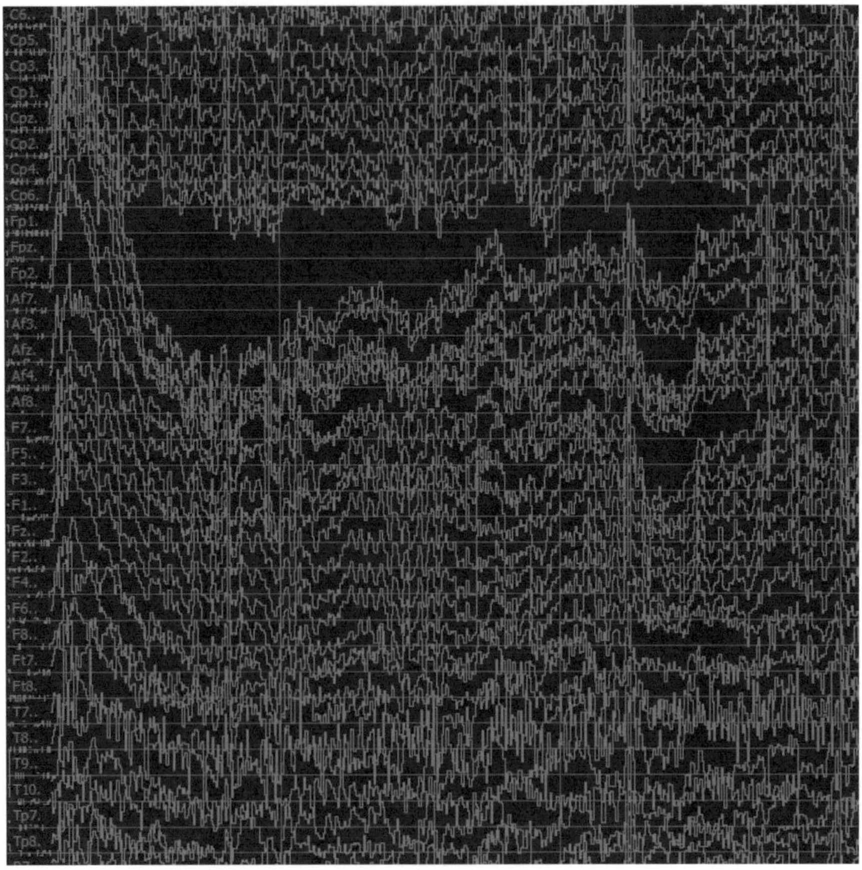

Figure 1.
Excerpt from a normal encephalogram. Channels on the left and brain waves monitored.

As a result of technological advances in both software and hardware, concepts such as artificial intelligence, machine learning or deep learning have been developed in last decades. This has allowed an evolution in many fields of society. In the field of brain signals, pattern recognition is of vital importance, both for diagnosis and for the development of applications that improve quality of life or simply for the development of mind-controlled tools or games. Thus, the Brain Computer Interfaces were born.

A brain-computer interface (BCI) or Brain-machine Interface is a system based on the recording or acquisition of the brain signal that is linked by direct communication (wired or wireless) to a machine or computer capable of interpreting and transforming thoughts or intentions into actual actions [7]. **Figure 2** shows the complete process of a BCI. First the obtaining of brain signals and their processing. Subsequently, the machine associated with the system is capable of receiving and interpreting these signals and resulting in a response.

In this way, using a public database, called Physionet, which consists of 109 subjects who perform tests of motor intentions, we are working on a BCI system whose objective is to be able to recognize the patterns of these thought movements and transfer them to a robotic arm that moves accordingly. Contributing to the current state of the art a new method that seeks to take into account the immediate previous mental state (IPMS) of each subject to improve pattern recognition.

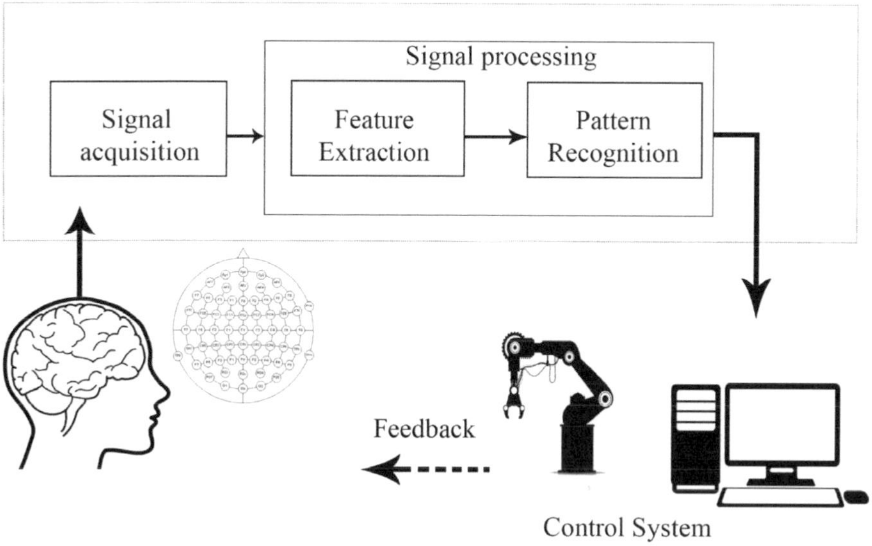

Figure 2.
Brain computer Interface workflow.

2. Pattern recognition with immediate previous mental state

2.1 State of art

From the last 10–15 years and until 2022, many studies have led to great advances in pattern recognition, human-computer interaction and in applications that use task classification or event-related potential (P300). Machine and deep learning algorithms and techniques have developed as computers and systems have improved over the years. They use different characteristics extracted from the data obtained from brain waves to classify or recognize certain patterns, getting better and better success in the pattern recognition and prediction [8–11]. E.g., in 2018 by Bird et al. made predictions of mental states of relaxation, neutrality and concentration [12]. Also, in emotional mental states such as being negative, neutral and positive [13].

In the field of EEG, large amounts of studies have been carried out in those years, addressing issues such as EEG channel selection, methods for make an optimal feature extraction or types of classifiers to predict patterns. For example, Feng et al. [14] published in 2019 an optimized channel selection method based on multi-frequency CSP-Rank for BCI systems using motor images. Jiménez et al. [15], propose an upper limb to assist in the rehabilitation of people with cerebrovascular accident or people with some disability or amputation. There are even lines of research that began to focus on the creation of brain-machine systems capable of acting under the orders of thought.

In 2020, philanthropist Elon Musk and his company Neuralink created an implantable BCI system to control systems with the mind which was successfully implanted in a pig in 2021. In addition, he announced that a monkey had been successfully made to play video games using this device [16]. The downside is that these BCI devices are invasive and still in the early stages of development.

All these advances always depend on the nature of the problem. In many cases the EEG image is taken directly and by using Common Spatial Patterns (CSP) the problem is solved. Other times 3D or 2D matrices are created to evaluate the problem and applying deep learning methods such as the use of Convolutional Neural Networks (CNN) together with Long-Short Term Memory (LSTM) the problem is solved.

The advantage of these techniques is the automation and success of the learning process. But, on the other hand, the disadvantage is the cost in terms of computation and the large amount of data that is required.

Normally all systems have work using a common scheme, as shown in **Figure 2**, which has the following steps:

1. Signal acquisition. Directly/EEG database.

2. Feature extraction.

3. Method selection for pattern recognition.

4. Train-Validation-Test. Feedback.

Subsequently, the rank of success in brain task classification of many works and also the most important work to date is shown.

In this chapter, a large number of avant-garde articles have been reviewed to extract the most important concepts and ideas in this field in order to adequately explain and expose the work carried out.

2.2 Pattern recognition with immediate previous mental state

As stated above, current works use machine learning and deep learning to recognize brain patterns and classify them. The success rate vary depends on the database used, the signal processing, the feature extraction or the types of classifiers used. Many of these works are between 61 and 76% in their success rate [17, 18].

The best work to date is that provided by Zhang et al. [19] where they claim to achieve success rates in the classification of at least 93% and up to 98%, using the same database (Physionet). They do this through cascade deep learning techniques, mixing a 3-layer CNN with recurrent neural networks (RNN) to take into account spatio-temporal characteristics of the experiments.

In other hand, we have a previous work, that are pending to be published, where we achieved experiments with up to 93% success using a mixture of machine learning classifiers. Taking this into account, we have done a lecture of the state of the art and concluded that according to several studies, such as that of Roc et al. [9], the success rate of the classification of mental tasks by in BCI systems is directly related to the subject in question and also to his mental state at that moment (relaxed, altered, nervous).

2.2.1 Our proposal

For this reason, our proposal consists of following a generalized scheme where the signal from the database is acquired, processed, the features are extracted and then a classification is done. However, we have decided to add the mental state prior to the moment of decision of the subjects as a variable, see **Figure 3**. That is, the moment before making a decision. To do this, after make a signal processing and a feature

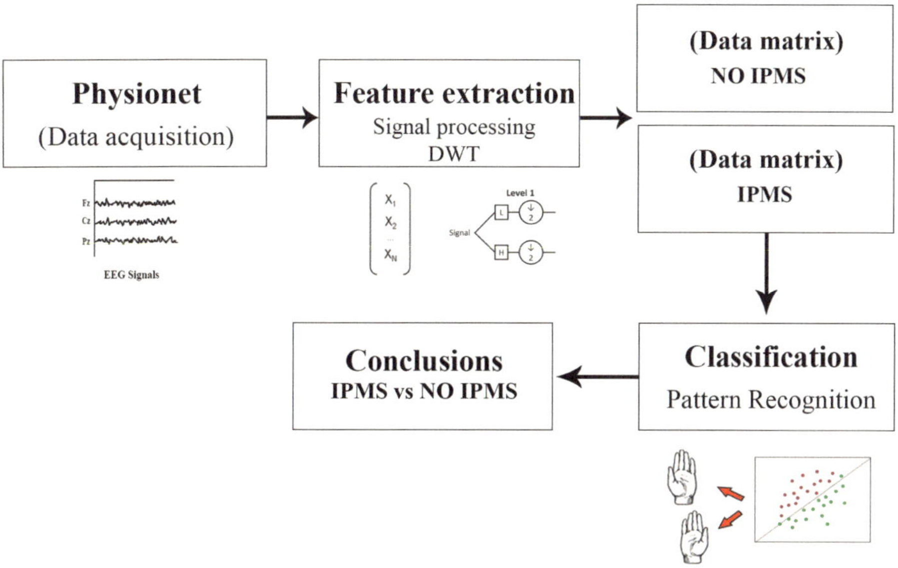

Figure 3.
Proposal workflow scheme.

extraction a classification of the mental tasks of various subjects with several classifiers of machine learning is done.

Later this is repeated, but taking into account their IPMS to compare the results. It seeks to demonstrate that there is a significant improvement in pattern recognition taking into account the average rank of success in classification and the standard deviation. We expect that the success rate in classification will increase and the standard deviation to decrease. This study is focus not in the success rate per se, but in the difference between taking into account IPMS or not.

2.3 Development of the BCI system

2.3.1 Signal acquisition

To carry out this work, the signal has been acquired by downloading the public database Physionet.org. A database developed in collaboration with the developers of the BCI 2000 system [20–22]. The database consists of 109 users who perform different types of tests, obtaining more than 1500 EEG records. Using a 10/20 system and placing a total of 64 electrodes. The tests to be carried out consist of 14 tests of approximately 2 minutes per test, where the subjects alternate periods of 4.1 s of rest with different tasks (of between 4.1 and 4.2 s) such as opening and closing their fist or imagining these movements.

In order to prove the hypothesis, in this chapter, several simplifications have been done:

- First. A group of 10 subjects is chosen as representative group of the set to demonstrate if there is evidence of success in having taken the IPMS into account.

- Second. Only experiments involving actual and thought movement of the left hand and right hand will be dealt with. We believe that if the hypothesis is proven here, it can be concluded that the same thing happens in the rest of the experiments.

- Third. According to studies, such as the one presented by Craik, A et al. [23], the motor activity associated with the hands is reflected in the frontal cortex of the brain and therefore the channels that contain major information are the 12 channels FC1, FC2, FC4, FC6, FCz, C1, C2 C3, C4, C5, C6 and Cz.

- Finally. The nomenclature that these experiments follow will be T0 for the rest intervals. T1 for the real or imaginary movement of the left hand and T2 for the real or imaginary movement of the right hand.

Obtaining in this way labeled data matrices for the experiments having taken into account the IPMS and without taking it into account. To obtain the IPMS, a difference between the interaction intervals (T1/T2) and the rest intervals prior to the motor imagery (T0) is made.

2.3.2 Feature extraction and pattern recognition

After acquiring the signal, the next step when developing a BCI system is to perform an optimal feature extraction that allows good pattern recognition.

The first thing to do is remove the noise that masks the signal and thus obtain a better signal-to-noise ratio. This is because when taking biological measurements, factors such as breathing itself, the heartbeat (low frequencies) or the electricity that runs through the circuit (high frequencies) are factors that add noise and can mask the signal. Considering the frequency of brain signals (1–100 Hz), in the EEG a key factor is to remove these unnecessary frequencies. Therefore, based on multiple studies [24–26], a bandpass filter is applied between 0.5 Hz and 50 HZ in order to avoid the electrical 50 Hz band and the low frequencies.

After noise removal, the discrete wavelet transform (DWT) is used. DWT is a signal processing tool used to perform multi-resolution analysis with variable time windows. It is a tool capable of decomposing and recomposing signals according to their time and frequency to facilitate the analysis [27]. Brain signals have unpredictable frequency and intensity over time, so they are non-stationary. Using the DWT is capable of breaking a signal with low-pass and high-pass filters at different levels, see **Figure 4**. Thus, obtaining a high frequency component and a low frequency component (with different information) at various levels. These levels correspond to different types of brain waves (delta to beta) that provide different types of information and could facilitate the pattern recognition.

Mathematically, the equation behind the DWT is given by Eq. (1), [28]:

$$f(t) = \sum_{k}\sum_{j} a_{j,k}\, \varphi_{j,k}(t) \tag{1}$$

This equation is expressed in terms of two indices, the translation time k and the scaling index j. These two indices are integers values and the wavelet functions form an orthogonal set of functions (base) [29].

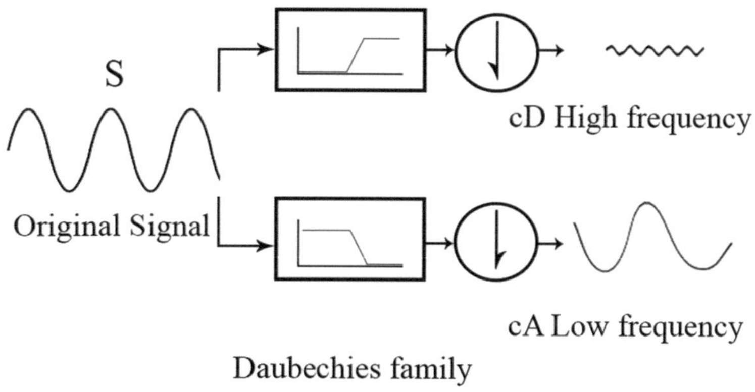

Figure 4.
Level 1 of decomposition using DWT. Original signal is separated in low and high frequency.

Daubechies family of wavelet have a better performance at time of classification as is shown by Alomari et al. [30] and others [20, 21].

Feature extraction allows at this point to have a better signal-to-noise ratio in the labeled matrices obtained in Section 2.1.1 and also a separation based on their frequencies. Thus, allowing pattern recognition when applying machine learning to be less expensive in terms of computing. This is due motor activities requires an active state of mind, and in consequence the result are the alpha and beta waves (8–30 Hz) already separated.

2.3.3 Classification

A series of machine learning classifiers are used to recognize the patterns of motor imagery (MI) and classify them. Each of the users has 3 real experiments and 3 MI experiments.

To test the hypothesis, the classifiers will be used with the MI experiments. In this way, it will be observed if there is an improvement in performance when taking into account the IPMS. The hold-out cross validation method is used to train and test the success in classification, training each subject with 70% of the set and blindly testing with 30%. The classifiers used are:

- Decision trees (DT) [31]
- Linear Discriminant Analysis (LDA) [32]
- Logistics Analysis (LA) [33]
- Support Vector Machine (SVM) [34]
- K-Nearest Neighbors (KNN) [35, 36]
- Ensemble methods (EM) [37]
- Neural networks (NN) [38]

We will not go into the mathematical details behind these classifiers as they are widely referenced and used in the world of pattern recognition and artificial intelligence. Attached is the reference where this information can be found.

3. Results and discussion

As explained in Section 2.2.1 the method to evaluate the success when taking into account the IPMS is the average success of the results and its standard deviation. In **Table 1**, the classify results without taking into account IPMS:

Users	DT	LDA	LA	SVM	KNN	EM	NN
S001_exp1_1	33.3	40.0	40.0	60.0	60.0	46.7	60.0
S001_exp1_2	53.3	40.0	40.0	53.3	80.0	66.7	26.7
S001_exp1_3	66.7	53.3	73.3	73.3	73.3	86.7	46.7
S002_exp1_1	66.7	66.7	73.3	80.0	80.0	73.3	60.0
S002_exp1_2	80.0	66.7	46.7	66.7	80.0	80.0	66.7
S002_exp1_3	53.3	33.3	60.0	53.3	86.7	60.0	53.3
S003_exp1_1	20.0	40.0	53.3	60.0	53.3	53.3	26.7
S003_exp1_2	26.0	60.0	40.0	73.3	73.3	66.7	73.3
S003_exp1_3	33.3	60.0	53.3	53.3	66.7	60.0	53.3
S004_exp1_1	26.7	33.3	93.3	53.3	53.3	33.3	53.3
S004_exp1_2	66.7	73.3	66.7	73.3	80.0	73.3	73.3
S004_exp1_3	40.0	53.3	53.3	60.0	73.3	60.0	60.0
S005_exp1_1	73.3	60.0	60.0	93.3	80.0	73.3	26.7
S005_exp1_2	26.7	26.7	46.7	60.0	60.0	53.3	46.7
S005_exp1_3	53.3	60.0	53.3	73.3	73.3	60.0	46.7
S006_cxp1_1	26.7	53.3	46.7	53.3	46.7	46.7	53.3
S006_exp1_2	53.3	40.0	86.7	53.3	80.0	66.7	66.7
S006_exp1_3	60.0	46.7	20.0	53.3	66.7	66.7	33.3
S007_exp1_1	40.0	53.3	60.0	60.0	73.3	73.3	73.3
S007_exp1_2	73.3	66.7	60.0	80.0	86.7	73.3	60.0
S007_exp1_3	40.0	33.3	40.0	53.3	60.0	53.3	26.7
S008_exp1_1	80.0	60.0	80.0	73.3	93.3	80.0	46.7
S008_exp1_2	13.3	46.7	86.7	66.7	53.3	53.3	60.0
S008_exp1_3	33.3	46.7	20.0	53.3	73.3	26.7	66.7
S009_exp1_1	66.7	46.7	66.7	53.3	66.7	73.3	53.3
S009_exp1_2	60.0	46.7	53.3	53.3	60.0	66.7	26.7
S009_exp1_3	73.3	46.7	40.0	53.3	73.3	73.3	73.3
S010_exp1_1	86.7	53.3	40.0	73.3	60.0	86.7	33.3
S010_exp1_2	40.0	33.3	46.7	53.3	53.3	53.3	33.3
S010_exp1_3	40.0	26.7	40.0	53.3	53.3	46.7	33.3

Table 1.
Classification without taking into account IPMS hypothesis.

In **Table 2**, the classify results taking into account IPMS:

It must be remembered that what is sought is to observe an improvement in pattern recognition by implementing the IPMS hypothesis and we are not focused on the search for the best classifier.

After obtaining the results of the classifications with and without IPMS, a comparison of the average of both methods is made, which is shown in **Table 3**. It can be seen that in 100% of the cases the classification improves significantly, with results of

Users	DT	LDA	LA	SVM	KNN	EM	NN
S001_exp1_11 1	40.0	33.3	53.3	66.7	53.3	53.3	60.0
S001_exp1_2	73.3	66.7	80.0	73.3	73.3	86.7	60.0
S001_exp1_3	53.3	33.3	73.3	60.0	66.7	73.3	33.3
S002_exp1_1	33.3	86.7	86.7	53.3	53.3	46.7	33.3
S002_exp1_2	73.3	60.0	46.7	86.7	86.7	73.3	46.7
S002_exp1_3	40.0	40.0	46.7	53.3	66.7	66.7	46.7
S003_exp1_1	46.7	46.7	20.0	53.3	73.3	53.3	60.0
S003_exp1_2	93.3	66.7	73.3	73.3	80.0	93.3	73.3
S003_exp1_3	80.0	46.7	60.0	60.0	60.0	80.0	40.0
S004_exp1_1	86.7	53.3	46.7	53.3	60.0	73.3	60.0
S004_exp1_2	60.0	46.7	53.3	53.3	60.0	60.0	53.3
S004_exp1_3	93.3	66.7	53.3	66.7	80.0	93.3	66.7
S005_exp1_1	60.0	60.0	53.3	66.7	73.3	53.3	46.7
S005_exp1_2	73.3	80.0	73.3	80.0	80.0	73.3	60.0
S005_exp1_3	40.0	40.0	40.0	53.3	53.3	46.7	33.3
S006_exp1_1	73.3	60.0	66.7	66.7	60.0	60.0	46.7
S006_exp1_2	80.0	73.3	80.0	80.0	80.0	86.7	80.0
S006_exp1_3	40.0	53.3	53.3	66.7	66.7	66.7	40.0
S007_exp1_1	60.0	46.7	66.7	53.3	86.7	60.0	46.7
S007_exp1_2	46.7	33.3	53.3	53.3	46.7	60.0	40.0
S007_exp1_3	66.7	40.0	46.7	60.0	86.7	86.7	66.7
S008_exp1_1	80.0	46.7	60.0	60.0	66.7	60.0	46.7
S008_exp1_2	73.3	73.3	93.3	86.7	80.0	93.3	40.0
S008_exp1_3	53.3	53.3	46.7	60.0	86.7	66.7	60.0
S009_exp1_1	73.3	60.0	66.7	73.3	73.3	66.7	53.3
S009_exp1_2	60.0	46.7	40.0	53.3	66.7	60.0	66.7
S009_exp1_3	60.0	33.3	40.0	53.3	66.7	60.0	46.7
S010_exp1_1	53.3	53.3	33.3	53.3	86.7	60.0	46.7
S010_exp1_2	86.7	46.7	53.3	60.0	73.3	73.3	60.0
S010_exp1_3	40.0	26.7	73.3	53.3	53.3	46.7	33.3

Table 2.
Classification taking into account IPMS hypothesis.

NO_IPMS	DT	LDA	LA	SVM	KNN	EM	NN
Avg_S001	51.1	44.4	51.1	62.2	71.1	66.7	44.5
Avg_S002	66.7	55.6	60.0	66.7	82.2	71.1	60.0
Avg_S003	26.4	53.3	48.9	62.2	64.4	60.0	51.1
Avg_S004	44.5	53.3	71.1	62.2	68.9	55.5	62.2
Avg_S005	51.1	48.9	53.3	75.5	71.1	62.2	40.0
Avg_S006	46.7	46.7	51.1	53.3	64.5	60.0	51.1
Avg_S007	51.1	51.1	53.3	64.4	73.3	66.6	53.3
Avg_S008	42.2	51.1	62.2	64.4	73.3	53.3	57.8
Avg_S009	66.7	46.7	53.3	53.3	66.7	71.1	51.1
Avg_S010	55.6	37.8	42.2	60.0	55.5	62.2	33.3
Total_Avg	51.1	48.6	53.8	62.2	68.6	63.9	49.6
STD	11.2	5.0	7.6	6.1	6.7	5.7	8.6
WITH_IPMS	DT	LDA	LA	SVM	KNN	EM	NN
Avg_S001	55.5	44.4	68.9	66.7	64.4	71.1	51.1
Avg_S002	48.9	62.2	60.0	64.4	68.9	62.2	42.2
Avg_S003	73.3	53.4	51.1	62.2	71.1	75.5	57.8
Avg_S004	80.0	55.6	51.1	57.8	66.7	75.5	60.0
Avg_S005	57.8	60.0	55.5	66.7	68.9	57.8	46.7
Avg_S006	64.4	62.2	66.7	71.1	68.9	71.1	55.6
Avg_S007	57.8	40.0	55.6	55.5	73.4	68.9	51.1
Avg_S008	68.9	57.8	66.7	68.9	77.8	73.3	48.9
Avg_S009	64.4	46.7	48.9	60.0	68.9	62.2	55.6
Avg_S010	60.0	42.2	53.3	55.5	71.1	60.0	46.7
Total_Avg	63.1	52.4	57.8	62.9	70.0	67.8	51.6
STD	8.7	8.0	7.0	5.3	3.5	6.3	5.3
Success rate with IPMS	DT	LDA	LA	SVM	KNN	EM	NN
Avg_Change (%)	12.0	3.8	4.0	0.7	1.4	3.9	2.0
STD_Change (%)	2.5	−3.0	0.6	0.8	3.2	−0.6	3.2

Table 3.
Comparison between NO IPMS and IPMS classification. Avg change and standard deviation changes of both methods.

up to 12% of improvement in classification. This leads us to think that for the recognition of brain patterns, taking into account the mental state prior to motor imagery is essential to obtain a better performance.

4. Conclusions

In this chapter a brief introduction to the field of brain signals has been made. The pattern recognition for the development of applications in fields such as medicine or industry and how to analyze brain signals for that propose has been explained. Subsequently, the BCIs have been introduced, explaining their operation and purpose. The IPMS hypothesis is proposed as an improvement of pattern recognition in BCI systems.

A public database (Physionet), where EEG records of subjects performing a series of motor and imaginary tasks are collected, is presented and using this dataset the steps to develop a BCI system are proposed. The signal is processed, features are extracted for pattern recognition and finally a series of classifiers are proposed to test the IPMS theory.

The results show there is evidence that taking into account the mental state prior to performing mental tasks directly affects the recognition of brain patterns and consequently the success in classifying them, improving them by up to 12%.

Note that research has been focused on testing this hypothesis and not on finding the best classifier. In future lines, we will try to apply the IPMS hypothesis to the best state-of-the-art classifiers.

Acknowledgements

This work was funded "Agencia Canaria de Investigación, Innovación y Sociedad de la Información de la Consejería de Economía Conocimiento y Empleo y por el Fondo Social Europeo (FSE) Programa Operativo Integrado de Canarias 2014-2020, Eje 3 Tema Prioritario 74 (85%)" from "Gobierno de Canarias" in Spain, under the reference "TESIS2020010118".

Appendices and nomenclature

BCI	brain computer interface
EEG	electroencephalogram
IPMS	immediate previous mental state
CSP	common spatial patterns
CNN	Convolutional Neural Network
LSTM	Long-Short Term Memory
RNN	Recurrent Neural Network
MI	Motor Imagery
DT	Decision Trees
LDA	Linear Discriminant Analysis
LA	Logistic Analysis
SVM	Support Vector Machine
KNN	K-Nearest Neighbors
EM	ensemble methods
NN	neural networks

Author details

Nabil Ajali-Hernández* and Carlos M. Travieso-Gonzalez
University of Las Palmas de Gran Canaria (ULPGC), Las Palmas de Gran Canaria, Spain

*Address all correspondence to: nabil.ajali101@alu.ulpgc.es

References

[1] CogniFit. CogniFit.com [Internet]. 2019. Available from: https://www.cognifit.com/es/cerebro

[2] Corbetta M, Shulman GL. Control of goal-directed and stimulus-driven attention in the brain. Nature Reviews. Neuroscience. 2002;**3**(3):201-215

[3] Hammond C. Cellular and Molecular Neurobiology. Deluxe ed. San Diego: Academic Press; 2001

[4] Buzsaki G. Rhythms of the Brain. Oxford: Oxford University Press; 2006

[5] Ramos-Argüelles F, Morales G, Egozcue S, Pabón RM, Alonso MT. Técnicas básicas de electroencefalografía: principios y aplicaciones clínicas. Spain: Canales del sistema sanitario de Navarra; 2009. pp. 69-82

[6] Barros MIM, Guardiola GT. Conceptos básicos de electroencefalograf {\'\i}a. Duazary. 2006;**3**(1):18-23

[7] Krucoff MO, Rahimpour S, Slutzky MW, Edgerton VR, Turner DA. Enhancing nervous system recovery through neurobiologics, neural interface training, and neurorehabilitation. Frontiers in Neuroscience. 2016;**10**:584

[8] Gajic D, Djurovic Z, Di Gennaro S, Gustafsson F. Classification of EEG signals for detection of epileptic seizures based on wavelets and statistical pattern recognition. Biomedical Engineering Applications Basis and Communications. 2014;**26**(02):1450021

[9] Oh SL, Hagiwara Y, Raghavendra U, Yuvaraj R, Arunkumar N, Murugappan M, et al. A deep learning approach for Parkinson's disease diagnosis from EEG signals. Neural Computing and Applications. 2020; **32**(15):10927-10933

[10] Fan M, Yang AC, Fuh J-L, Chou C-A. Topological pattern recognition of severe Alzheimer's disease via regularized supervised learning of EEG complexity. Frontiers in Neuroscience. 2018;**685**:3-5

[11] Gurumoorthy S, Muppalaneni NB, Gao X-Z. Analysis of EEG to find Alzheimer's disease using intelligent techniques. In: Computational Intelligence Techniques in Diagnosis of Brain Diseases. Singapore: Springer; 2018. pp. 61-70

[12] Bird JJ, Manso LJ, Ribeiro EP, Ekart A, Faria DR. A study on mental state classification using eeg-based brain-machine interface. In: International Conference on Intelligent Systems (IS). Madeira: IEEE; 2018. pp. 795-800

[13] Bird JJ, Ekart A, Buckingham CD, Faria DR. Mental emotional sentiment classification with an eeg-based brain-machine interface. In: Proceedings of the International Conference on Digital Image and Signal Processing (DISP'19). Oxford: 2019

[14] Feng JK, Jin J, Daly I, Zhou J, Niu Y, Wang X, et al. An optimized channel selection method based on multifrequency CSP-rank for motor imagery-based BCI system. Computational Intelligence and Neuroscience. 2019:6-8

[15] Jiménez AR, Grisales CA, Sotelo JL. Diseño de un sistema cerebro-maquina de miembro superior para la asistencia a la rehabilitación de personas con accidente cerebro-vascular. Encuentro Int Educ en Ing. 2019

[16] Pérez E. Un mono jugando al Pong es la primera demostración de Neuralink, el

proyecto de Elon Musk para conectar el cerebro con los ordenadores [Internet]. Xakata. 2021. Available from: https://www.xataka.com/investigacion/mono-jugando-al-pong-primera de mostracion-neuralink-proyecto-para-conectar-cerebro-ordenadores-elon-musk

[17] Major TC, Conrad JM. The effects of pre-filtering and individualizing components for electroencephalography neural network classification. Southeast Construction. 2017;**2017**:1-6

[18] Wu SL, Liu YT, Hsieh TY, Lin YY, Chen CY, Chuang CH, et al. Fuzzy integral with particle swarm optimization for a motor-imagery-based brain–computer interface. IEEE Transactions on Fuzzy Systems. 2016; **25**(1):21-28

[19] Zhang D, Yao L, Chen K, Wang S, Chang X, Liu Y. Making sense of spatio-temporal preserving representations for EEG-based human intention recognition. IEEE Transactions on Cybernetics. 2019; **50**(7):3033-3044

[20] Physionet.org. PhysioNet [Internet]. 2019. Available from: https://www.physionet.org/physiobank/database/eegmmidb

[21] Goldberger AL, Amaral LAN, Glass L, Hausdorff JM, Ivanov PC, Mark RG, et al. PhysioBank, PhysioToolkit, and PhysioNet: Components of a new research resource for complex physiologic signals. Circulation. 2000;**101**(23):215-220

[22] Schalk G, McFarland DJ, Hinterberger T, Birbaumer N, Wolpaw JR. BCI2000: A general-purpose brain-computer interface (BCI) system. IEEE Transactions on Biomedical Engineering. 2004;**51**(6):1034-1043

[23] Craik A, He Y, Contreras-Vidal JL. Deep learning for electroencephalogram (EEG) classification tasks: A review. Journal of Neural Engineering. 2019; **16**(3):31001

[24] Ji N, Ma L, Dong H, Zhang X. EEG signals feature extraction based on DWT and EMD combined with approximate entropy. Brain Sciences. 2019;**9**(8):201

[25] Medina B, SIERRA JE, ULLOA AB. Técnicas de extracción de caracter{\'\i}sticas de señales EEG en la imaginación de movimiento para sistemas BCI. Revista ESPACIOS. 2018:8-10

[26] Noguera MAP, Ortega CEM, Castro W, Ordoñez DH. Análisis De Señales EEG Para Detección De Intenciones Motoras Aplicadas A Sistemas BCI.

[27] Bhattacharya S, Haddad RJ, Ahad M. A Multiuser EEG Based Imaginary Motion Classification Using Neural Networks. Southeastern Construction. Norfolk: IEEE; 2016. pp. 1-5

[28] Burrus CS. Introduction to Wavelets and Wavelet Transforms: A Primer. Englewood Cliffs. New Jersey: Prentice Hall; 1997

[29] Wei D, Tian J, Wells RO, Burrus CS. A new class of biorthogonal wavelet systems for image transform coding. IEEE Transactions on Image Processing. 1998;**7**(7):1000-1013

[30] Alomari MH, AbuBaker A, Turani A, Baniyounes AM, Manasreh A. EEG mouse: A machine learning-based brain computer interface. International Journal of Advanced Computer Science and Applications. 2014;**5**(4):193-198

[31] Strobl C, Malley J, Tutz G. An introduction to recursive partitioning: Rationale, application, and characteristics of classification and

regression trees, bagging, and random forests. Psychological Methods. 2009; **14**(4):323

[32] McLachlan GJ. Discriminant Analysis and Statistical Pattern Recognition. New Jersey: John Wiley & Sons; 2005

[33] Cramer JS. The Origins of Logistic Regression. UK: Cambridge University Press; 2002

[34] Cortes C, Vapnik V. Support-vector networks. Machine Learning. 1995; **20**(3):273-297

[35] Altman NS. An introduction to kernel and nearest-neighbor nonparametric regression. American Statististics. 1992;**46**(3):175-185

[36] Sayad DS. An Introduction to Data Science [Internet]. 2021. Available from: https://injuryfacts.nsc.org/motor-vehicle/overview/introduction/

[37] Natekin A, Knoll A. Gradient boosting machines, a tutorial. Frontiers in Neurorobotics. 2013;7:21

[38] Lv C, Xing Y, Zhang J, Na X, Li Y, Liu T, et al. Levenberg–Marquardt backpropagation training of multilayer neural networks for state estimation of a safety-critical cyber-physical system. IEEE Transactions on Industrial Informatics. 2017;**14**(8):3436-3446

Chapter 8

Multi-Features Assisted Age Invariant Face Recognition and Retrieval Using CNN with Scale Invariant Heat Kernel Signature

Kishore Kumar Kamarajugadda and Movva Pavani

Abstract

Face recognition across aging emerges as a significant area among researchers due to its applications such as law enforcement, security. However, matching human faces with different age gaps is still bottleneck due to face appearance variations caused by aging process. In regard to mitigate such inconsistency, this chapter offers five sequential processes that are Image Quality Evaluation (IQE), Preprocessing, Pose Normalization, Feature Extraction and Fusion, and Feature Recognition and Retrieval. Primarily, our method performs IQE process in order to evaluate the quality of image and thus increases the performance of our Age Invariant Face Recognition (AIFR). In preprocessing, we carried out two processes that are Illumination Normalization and Noise Removal that have resulted in high accuracy in face recognition. Feature extraction adopts two descriptors such as Convolutional Neural Network (CNN) and Scale Invariant Heat Kernel Signature (SIHKS). CNN extracts texture feature, and SIHKS extracts shape and demographic features. These features plays vital role in improving accuracy of AIFR and retrieval. Feature fusion is established using Canonical Correlation Analysis (CCA) algorithm. Our work utilizes Support Vector Machine (SVM) to recognize and retrieve images. We implement these processes in FG-NET database using MATLAB2017b tool. At last, we validate performance of our work using seven performance metrics that are Accuracy, Recall, Rank-1 Score, Precision, F-Score, Recognition rate and computation time.

Keywords: age-invariant face recognition, image quality evaluation, pose normalization, multiple feature extraction, recognition and retrieval

1. Introduction

As one of the most significant topics in computer vision and pattern recognition, face recognition attains much attention from both academic and industries over recent decades [1, 2]. With the evolution of neural networks, general face recognition technology emerged as a noteworthy area among researchers [3–5]. However, identifying face images across widespread range of ages is shortcoming due to human face appearance changes affected by aging process [6, 7]. In order to achieve human face recognition under difference ages, Age-Invariant Face Recognition (AIFR) approach is progressed [8]. AIFR recognizes faces using facial features extracted from human images. AIFR method uses three different models such as generative, discriminative [9], and deep learning methods [10]. Generative approaches are based on the age progression methods in regard to converting the probe image into the same age as that of gallery image [11]. However, generative schemes have several shortcomings [12]. Optimizing the recognition performance in generative model is not easier task. Estimating the accurate results in generative model is highly difficult since it cannot handle aging impact. Discriminative approaches [13] are introduced to resolve discrepancy of generative scheme [14]. It develops feature matching using local descriptors [15] in AIFR. Multiple descriptors-based AIFR is introduced to extract features from periocular region [16]. In this, two descriptors are used to extract features that are Scale-Invariant Feature Transform (SIFT) and Speeded-Up Robust Features (SURF).

In order to achieve better result in AIFR, deep learning method is integrated with discriminative approach [17]. In deep learning, Convolutional Neural Network (CNN) algorithm plays vital role in recognizing face with different aging images [18]. Large age gap verification is performed by injecting features in deep networks [19]. Here, deep CNN is used to recognize face where texture features are considered. Aging model-based face recognition with different aging images is introduced under deep learning method [20]. Here, CNN descriptor is utilized to match image with different aging images.

From the aforesaid studies, we determine that there are still many issues present in recognizing face with aging progress. The issues are discussed as follows:

- Preprocessing is not effective in most of the chapter that reduces performance of the system.
- Pose normalization is not considered in existing AIFR, which is highly significant. Since, AIFR datasets such as MORPH, FG-NET, etc., contain different pose images.
- Existing feature extraction procedures lack in extracting features from important regions that tend to reduce recognition rate.
- Face recognition algorithms are not up to the level to handle large dataset and thus reduce the result of accuracy.

These problems impose confines on the present AIFR systems and also complicate the recognition and retrieval task especially under different aging images.

1.1 Research contribution

In order to tackle abovementioned issues, our work contributes the following processes:

- In order to reduce time wastages in preprocessing, we initially execute novel Image Quality Evaluation (IQE) method, which estimates Image Quality Metric (IQM) for each image. If IQM value is below Image Quality Threshold (IQT), then only preprocessing is performed for that image or else directly gone into the pose normalization process.

- Preprocessing is performed to reduce uncertainties in upcoming face recognition processes such as feature extraction, recognition, and retrieval. For this purpose, we implement two processes such as illumination normalization and noise removal. Illumination normalization adopts DGC-CLAHE and noise removal adopts ASBF algorithm.

- Pose normalization is significant to diminish difficulties present in feature extraction and thus enhances the recognition and retrieval performance.

- Our work extracts feature from three regions that are periocular, nose, and mouth in order to increase recognition rate. Here, two descriptors are utilized that are CNN and SIHKS, which perform better than other existing descriptors such as LBP, SIFT, etc.

- In order to reduce recognition and retrieval time, we fuse features after extraction using CCA.

- Recognition and retrieval are performed through SVM algorithm, which performs well even with unstructured, semistructured data such as text, images, and trees.

1.2 Research outline

Outline of this chapter is summarized as follows: Section 2 deliberates state-of-the-art works existing in AIFR with their limitations. Section 3 exemplifies problems occurring in previous works related to AIFR. Section 4 explains brief study of our proposed work with our proposed algorithms. Section 5 illustrates numerical results obtain from our simulation environment and also compares it with existing methods. Finally, section 6 concludes our contribution and also provides comment on our future work.

2. Related work

This section discusses the state-of-the-art work related to AIFR along with their limitations. In this, we discussed works that comprise preprocessing, feature extraction, recognition, and retrieval processes.

Kishore et al. [21] have suggested Periocular Region-Based AIFR Using Local Binary Pattern. In this, three sequential processes are executed to recognize faces that are preprocessing, feature extraction, and classification. In preprocessing, enhancement and denoising processes are employed in each facial image. Local Binary Pattern (LBP) descriptor [22] was used to extract features from the periocular region [23] of the given face image. Periocular region contains eyes, eye lashes, and eye brow parts of the face. Chi-square distance was used as classifier to recognize face after feature extraction. Chi-square distance doesn't recognize face accurately since it is highly sensitive to the sample size.

Nanni et al. [24] have introduced ensemble of texture descriptor and preprocessing techniques to recognize image effectually. Four face recognition processes are performed that are preprocessing, feature extraction, feature transform, and classification. Preprocessing executes three techniques that are adaptive single index retinex (AR) in order to enhance scene detail and color enhancement in darker area. Anisotropic smoothing and different of Gaussian (DoG) are algorithms executed to normalize the illumination field. Features are extracted using two descriptors that are Patterns of the Oriented Edge Magnitudes (POEM) and the Monogenic Binary Coding (MBC). At last, different distance functions are used to recognize face. Accuracy of face recognition was very less due to poor feature extraction mechanism. Chi et al. [25] have offered temporal nonvolume preserving approach to facial age progression and AIFR. In preprocessing, face region was detected and aligned based on the fixed position of the eyes and mouth corners. And then it maps the texture features of the test image with the trained image in order to verify images. Here, deep CNN algorithm was utilized to map features. In this, preprocessing step doesn't perform effective processes such as normalization, noise removal that tend to reduce system performance.

Bor et al. [26] have introduced Cross Age Reference Coding (CARC) for AIFR. Initially, it executes face detection algorithm in order to detect face region in image. And it extracts features from the detected region for which it utilizes high-dimensional LBP algorithm. LBP extracts 59 local features from the detection regions. In this, Principal Component Analysis (PCA) algorithm was used to reduce dimensionality of extracted feature. After that, CARC recognizes face using local features transformation. More analysis is required on feature extraction since it plays vital role in AIFR. Yali et al. [27] have pointed out distance metric optimization driven CNN for AIFR. Here, two models are integrated that are feature learning and distance metric learning. This integration is achieved through CNN algorithm with parameters optimized using network propagation algorithm. CNN learns features using the convolution layer and recognizes face using the distance metric. Finally, recognized images are retrieved effectually. Herein, recognition rate was very less due to ineffective feature extraction.

Pournami et al. [28] have offered deep learning and multiclass SVM algorithm to recognize face. Here, preprocessing was performed to increase the accuracy of the face recognition where image resizing was performed. CNN feature descriptor was used to extract features from the given image. Here, fully connected layer extracts features from the image and then features are given as input to the multiclass SVM classifier. Resizing only performed in preprocessing thus introduced more noise in extracted feature. Garima et al. [29] have suggested techniques for face verification across different age progression with large age gap. Initially, image normalization was performed where RGB image was converted into the grayscale image and the image is rotate as the eyes are aligned horizontally. In this, face features are extracted using Center Symmetric Local binary Pattern (CSLBP) algorithm. And also weighted K-nearest Neighbor (K-NN) algorithm was used to recognize face from extracted features. K-NN doesn't perform well for large dataset and thus reduces the accuracy of face recognition. Saroj et al. [30] have pointed out pyramid binary pattern for age-invariant face verification. In this, pyramid binary pattern was used to extract texture feature. Texture features are given as input to the PCA in order to reduce dimensionality of the extracted features. And then, classification was performed through SVM algorithm. Here, texture feature was only extracted to classify the face with age invariant. Thus it reduces accuracy in face recognition since dataset contains different images with large age gap.

Mrudula et al. [31] have offered face recognition across aging using GLBP features. Preprocessing performs three sequential processes that are image resizing, RGB to gray, and illumination normalization. Here, combined feature descriptor was used to extract features from the given image. LBP and Gabor descriptors are combined, which was known as GLBP descriptor. During classification, PCA was used to reduce feature dimensionality and K-NN algorithm was used to recognize face across aging. Herein, GLBP descriptor introduces high false-positive rate in age-invariant face recognition. Zhen et al. [32] have pointed out local polynomial contrast binary patterns for face recognition. Polynomial filters are used to extract the attributes from the given image. In this, LBP descriptor was used to extract texture from the given image. Fisher Linear Discriminant (FLD) algorithm is used to reduce dimension of extracted features. Here, extracted features are classified using nearest neighbor classifier to recognize given image in training set. Nearest neighbor classifier consumes more time to classify image since all the work is performed in testing stages only.

Mohanraj et al. [33] have suggested ensemble of CNN for face recognition in order to resolve aging, pose variation, and low-resolution problem. Preprocessing was established to resize the given image. After that, features are extracted using three different CNN algorithms. Features are concatenated and given to the classifier in order to predict the person. Here, random forest classifier is used to recognize the face. Noise removal was not performed in preprocessing and thus reduces the accuracy of face recognition. Rupali et al. [34] have introduced component-based face recognition. Here, three face components are considered that are nose, lips, and ears. Preprocessing is performed to resize the image and features are extracted using CNN algorithm. Features are extracted from nose and face regions that are given to the FLD algorithm to reduce the dimensions. These features are given to KNN classifier in order to predict the image. In KNN, initial K value prediction is complex that leads to ineffective results. Venkata et al. [35] have pointed out real-time face recognition using deep learning and LBP. During preprocessing, it resizes the given image. In this, LBP was used to extract features from the given images. Extracted features are given to the CNN in order to provide weight to each feature. CNN provides weight in order to estimate the matched face with the training images. Here, texture feature only extracted to recognize face across aging that tends to reduce recognition rate.

Mohsen et al. [36] have offered age-based human face image retrieval using zernike moments. In this, Zernike moment was used to extract features from the images. Here, Zernike moment utilizes Zernike Basis Function (ZBF), which captures both local and global featured fro face image. And, Multi-Layer Perceptron (MLP) algorithm was used to recognize age in training image. Accurate result was not obtained in MLP classifier, thus reducing the recognition rate. Danbei et al. [37] have offered face aging synthesis application based on feature fusion. Initially, face detection was performed and feature points are positioned. For this purpose, triangulation and affine transformations are used, which position the feature points. Here, facial texture features are extracted to recognize face across aging. Extracted features are fused in order to recognize face with the training images effectually. More analysis is required on facial recognition since it describes up to feature fusion process.

3. Problem statement

Kishore et al. [38] have offered Hybrid Local Descriptor (HLD) and LDA-assisted K-Nearest Neighbor classification in AIFR. Here, Gaussian filter was used to reduce noise that results in information degradation, since it removes fine details of the image and resultant image is blurred. WLD-based feature extraction loses more information due to lack of pixel consideration. K-NN-based classification requires more time due to absence of training phase and finding good similarity measure is also difficult. Muhammad et al. [39] have introduced Demographic Features (DF)-assisted AIFR and retrieval. In this, feature extraction takes more time, since each feature was extracted in three individual CNNs. Position and orientation of the object were ignored in hidden layer of CNN that result in less accuracy in feature extraction and recognition. Chenfei et al. [40] have pointed out Coupled Auto Encoder (CAN) algorithm based feature extraction in AIFR. Herein, feature extraction was not effective due to lack of texture and shape-oriented features. In CAN, data relationships are not considered that affect classification results and weight computation is also very difficult. Huiling et al. [41] have introduced Identity Inference Model (IIM)-based age subspace learning to recognize image in AIFR. Herein, wLBP-based feature extraction was used that results in less accuracy, since it contains more noise in extracted features due to absence of noise removal process. Fahad et al. [42] have introduced Composite Temporal Spatio (CTS) modeling in order to recognize image in AIFR. Here, preprocessing was required to improve the accuracy in age-invariant face recognition, since image database contains illumination, pose variation, etc. Naïve Bayes–based classification results are always biased one, since it doesn't rely on class conditional dependency.

4. Proposed work

This section briefly describes our proposed method in detail along with the description of utilized algorithm.

4.1 System overview

Our Multi-Feature-assisted AIFR (MF-AIFR) method tackles problems that are present in the previous AIFR works. For this purpose, MF-AIFR establishes the five consecutive processes that are IQE, Preprocessing, Pose Normalization, Feature Extraction and Fusion, Feature Recognition and Retrieval as depicted in **Figure 1**. Our work novelty is present in the IQE method, since previous AIFR method doesn't concentrate on the quality evaluation. In order to save time, MF-AIFR performs IQE where images that are not satisfied IQT only given to the preprocessing step or else it is directly given to the pose normalization process. During preprocessing, MF-AIFR performs two processes that are illumination normalization using DGC-CLAHE and noise removal using ASBF algorithm. Pose normalization is executed to enhance feature extraction performance where EA-AT algorithm is utilized. Multiple features are extracted from the three different regions of face image that are periocular, mouth, and nose in regard to enhancing accuracy result. Here, two descriptors are executed that are CNN for texture feature and SIHKS for demographic and shape features extraction. Here, demographic features comprise age, gender, and race. Extracted features are fused using CCA in accord to reduce the complex recognition process. For recognition and retrieval, MF-AIFR pursues SVM algorithm, which has high scalability compared with other machine learning algorithm.

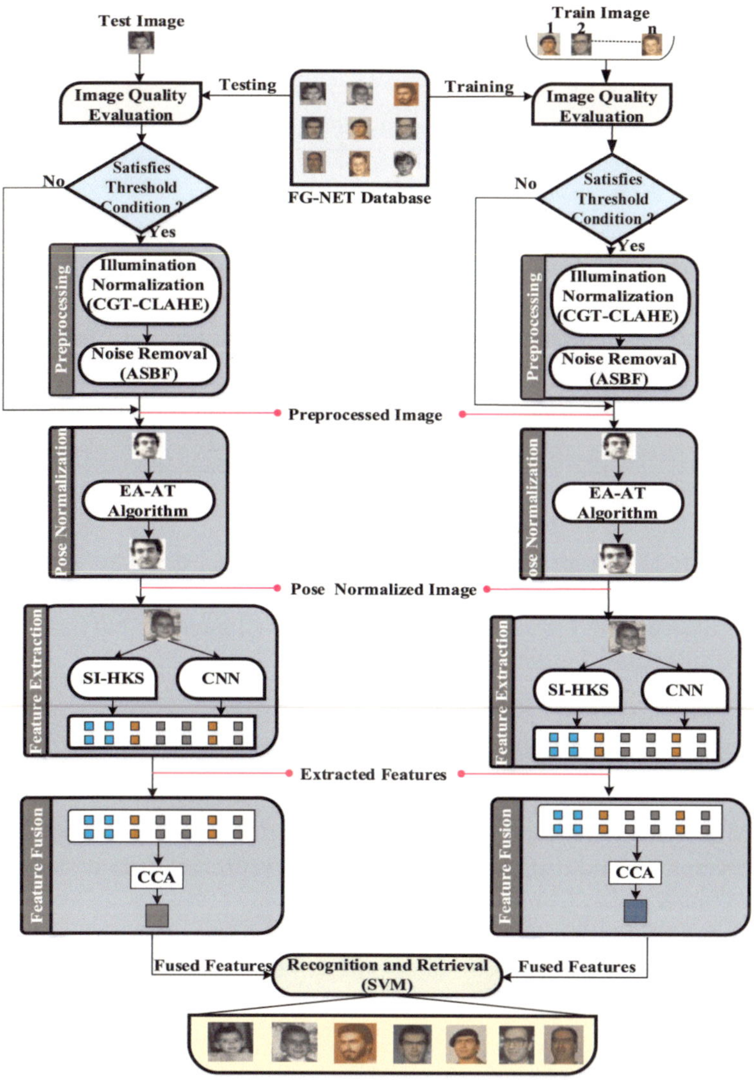

Figure 1.
Architecture for proposed work.

Figure 1 illustrates the architecture for our proposed work. The process depicted in architecture is described briefly in upcoming sections.

4.1.1 Image quality evaluation (IQE)

Reducing computation time in AIFR and retrieval is noteworthy in order to achieve efficient performance. For this purpose, MF-AIFR performs novel IQE, which estimates IQM for each image. IQM comprises subsequent metrics that are Brightness *Evaluation*, Region Contrast *Evaluation*, Edge Blur *Evaluation*, Color Quality *Evaluation*, and Noise *Evaluation*.

These metrics are designated as follows:

Brightness Evaluation: It defines the image darkness degree in order to get easy viewing. It can be measured as follows:

$$I_{BD} = \frac{\left\{\sum_{k_1=1}^{N_1}\sum_{k_2=0}^{255} h_{k_1k_2} \times (k_2)^s\right\}}{N} \tag{1}$$

Where $h_{k_1k_2}$ represents the pixel quantity of the gray value k_2 in the histogram of the k_1^{th} image block. s indicates the parameter where $s = 3$. N indicates number of sample blocks.

Region Contrast Evaluation: This metric is used to distinguish difference images effectually. It can be measured through below expression:

$$I_{CD} = \frac{\left\{\sum_{k=1}^{N_1}\left[I_k^{max} - I_k^{min}\right]/\left[I_k^{max} + I_k^{min}\right]\right\}}{N} \tag{2}$$

Where I_k^{max} and I_k^{min} represent the maximum and minimum gray values of the kth image block.

Edge Blur Evaluation: It defines the clearness of the image for easy analysis. This metric can be measured as follows:

$$I_{EBD} = \max_{I \in \odot} \frac{\{\arctan\left[I(i_1, j_1) - I(i_2, j_2)\right]\}}{Wid_{12}} \tag{3}$$

Where $\odot$ denotes set of image blocks, $I(i_1, j_1)$ and $I(i_2, j_2)$ represent the gray values of first and second image blocks. And, Wid_{12} represents the width of the edge spread points such as (i_1, j_1) and (i_2, j_2).

Color Quality Evaluation: This metric defines the image quality in terms of the color. It can be measured using following expression:

$$I_{CQD} = \frac{\sum_{i=1}^{l}\sigma_i^C}{l} \tag{4}$$

Where σ_i^C represents the ith standard deviation of the component intensity in the HSV color space. l indicates channel number, $l = 3$.

Noise Evaluation: This metric can be used to measure noise present in the image. It can be measured using following expression:

$$I_{ND} = \frac{\sigma_n}{I_{BD}} \tag{5}$$

Where σ_n represents the standard deviation of the image block. It can be estimated using below expression,

$$\sigma_n = a \times \lg\left(b \times \frac{255}{I_{BD}}\right) \times \min_{I \in \odot} \sigma_i \tag{6}$$

Where a and b are constant values.

Using above parameters, we estimate IQM for each image. It can be measured as follows:

$$IQM = \sum \frac{I_{BD} + I_{CQD} + I_{CD}}{I_{ND} + I_{EBD}} \tag{7}$$

After computing IQM, this value is compared with the IQT in order to select whether next process is preprocessing or pose normalization for given image.

$$Ne_{p,i} = \begin{cases} IQM_i > IQT \rightarrow Pose\ Normalization \\ IQM_i < IQT \rightarrow Preprocessing \end{cases} \tag{8}$$

Where $Ne_{p,i}$ represents the next process for image i. Using the above condition, we select next process for each given image and thus avoid time wastages in performing preprocessing for all images. In the meantime, it also reduces the total computation time for recognition and retrieval.

4.1.2 Preprocessing

MF-AIFR performs preprocessing in order to enhance the recognition rate in simulation results. For this purpose, we perform two processes in preprocessing that are illumination normalization and Noise removal.

4.1.2.1 Illumination normalization

Illumination normalization is performed in order to enhance the image quality and also avoid negative effects of the image. MF-AIFR adopts DGC-CLAHE algorithm for illumination normalization. Proposed DGC-CLAHE performs better than existing CLAHE method. It enhances both luminance and contrast of the image adaptively. Our DGC-CLAHE algorithm performs dual gamma correction, which enhances the dark areas of the image. This algorithm adaptively sets the clip points of each image, which depends on the dynamic range of each block of the image. In this, first gamma correction is executed to boost the entire luminance present in the image block. Second gamma correction is executed to adjust the contrast in very dark region in order to avoid overenhancement in bright regions.

Initially, DGC-CLAHE sets clip point adaptively based on the dynamic range, which can be expressed as follows:

$$\beta = \frac{p}{d_r}\left(1 + \tau\frac{g_{max}}{R} + \frac{\alpha}{100}\left(\frac{\sigma}{A_v + c}\right)\right) \tag{9}$$

Where p represents number of the pixels in each block, d_r denotes dynamic range in this block. τ and α represent the constant parameters, which are used to control the weight of dynamic range and entropies. σ indicates the standard deviation of the block; A_v indicates mean value; and c represents the small value in order to avoid division by 0. R represents entire dynamic range of the image. $g_{\max}$represents maximum pixel value of the image. After completing setting up of clip points, dual gamma corrections are performed.

DGC-CLAHE defines enhancement weight for the global gray levels of the blocks by first gamma correction (γ_1), which can be expressed as follows:

$$W_e = \left(\frac{Gr_{max}}{Gr_{ref}}\right)^{1-\gamma_1} \tag{10}$$

Where Gr_{max} indicates maximum gray value of the image, and Gr_{ref} indicates reference gray value of the image. First (γ_1) and second gamma (γ_2) corrections are represented as follows:

$$\gamma_1 = \frac{\ln\left(o + cdf_{\omega}(Gr_l)\right)}{8} \tag{11}$$

$$\gamma_2 = \frac{1 + cdf_{\omega}(Gr_l)}{2} \tag{12}$$

Where o indicates constant, cdf_{ω} indicates cumulative distribution function weight, and Gr_l represents gray level of the image. γ_1 and γ_2 are increased by Gr_l in order to avoid under enhancement in darker region of the image. The first and second gamma correction setting based normalization provides better result in image with nonuniform illumination. Thus, it enhances the image effectually, which in turn increases recognition rate.

4.1.2.2 Noise removal

Noise removal is substantial process in face recognition in regard to enhancing recognition accuracy. For this purpose, our MF-AIFR utilizes ASBF algorithm to remove noise from given image. Proposed ASBF algorithm preserves fine details of the image while removing noise and also sharpens the image. ASBF algorithm is used to remove universal noises such as impulse and Gaussian.

In ASBF algorithm, noisy pixel is detected using Sorted Quadrant Median Vector (SQMV), which incorporates significant features such as edge or texture information. Our ASBF algorithm executes three sequential processes as depicted in **Figure 2**. Initially, Adaptive Median Filter (AMF) is used to identify the corrupted pixels in the image. Secondly, the edge of the image is preserved using edge detector, which accurately predicts the edge existence in the current window. Noise detector is used to classify the noise into impulse and Gaussian. Switching Bilateral Filter (SBF) contains ranging filter, which switches the modes between impulse and Gaussian based on noise detector result.

4.1.2.2.1 AMF

Existing noise filtering algorithm utilizes constant window size such as 3*3, which may fail to distinguish noisy and noise-free pixel accurately and thus results in blur output image. In order to avoid this drawback, our AMF adaptively changes the window size based on the number of noisy pixels present in given image.

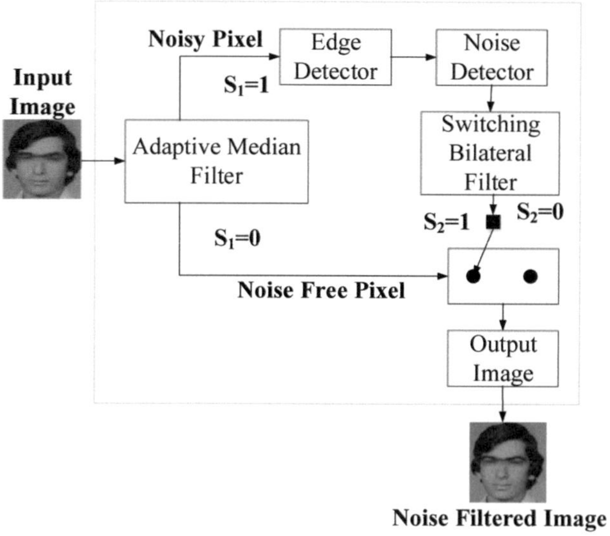

Figure 2.
ASBF function blocks.

4.1.2.2.2 Noise detector

Noise detector is used to predict whether pixel is filtered by SBF Gaussian (SBF_g) or SBF impulse (SBF_i). Let us consider S_1 and S_2, which are binary control signals where S_1 is generated by AMF and S_2 is generated by noise detector. Then the filtered image is represented as follows:

$$f_p = \begin{cases} SBF_g \; S_1 = 1^{S_2} = 1 \\ SBF_i \; S_1 = 1^{S_2} = 0 \\ n_{fp} \; S_1 = 0^{S_2} = 0 \end{cases} \tag{13}$$

At last, pixel with Gaussian and impulse noises are classified based on the above discussed conditions. These outputs are given as input to the SBF with SQMV.

SBF with SQMV: SBF switches its mode based on the classification results from the noise detector. Here, SQMV scheme is used to predict the optimum median effectively even in the larger window. SMQV detects noisy pixel by estimating difference between current pixel and reference median pixel. If difference is large, then current pixel is considered as the noisy pixel. Let us consider $\rho_{i,j}$ as the current pixel and $\rho_{i+s,j+t}$ as the pixels in a $(2N + 1) \times (2N + 1)$ window surrounding $\rho_{i,j}$.

The output from the SBF filter is expressed as follows:

$$O_{i,j} = \frac{\sum_{m=-n}^{n} \sum_{t=-n}^{n} W_{sr}(m,t) \rho_{i+s,j+t}}{\sum_{m=-n}^{n} \sum_{t=-n}^{n} W_g(m,t) W_{sr}(m,t)} \tag{14}$$

Where,

$$W_g = e^{-\frac{\left[(i-m)^2 + (j-t)^2\right]}{2\sigma_s^2}} \tag{15}$$

$$W_{sr} = e^{-\frac{I-\rho_{i+s,j+t}}{2\sigma_R^2}} \tag{16}$$

Where I represents the reference median for impulse noise ($S_1 = 1$ and $S_2 = 1$) and $I = \rho_{i,j}$ for Gaussian noise ($S_1 = 1$ and $S_2 = 0$).

From the above discussions, we conclude that our proposed ASBF removes not only Gaussian noise but also impulse noise while keeping the image fine details and images. This way of performing preprocessing increases the accuracy in AIFR.

4.1.2.3 Pose normalization

Pose normalization is substantial process to increase accuracy in face recognition. Since, our database FG-NET contains different pose images and thus requires pose normalization before entering into feature extraction and retrieval. Our MF-AIFR carried out EA-AT algorithm in order to correct the different poses into the frontal view and thus increases the feature extraction efficiency. EA-AT algorithm initially estimates pose angle of given image using Euler Angle. Then, estimated angle is provided to the Affine Transformation to get frontal view of the given image. Euler angles are three angles in order to describe the orientation of the face with respect to the fixed coordinate.

Figure 3 illustrates the Euler angle with their coordinates in Z vector. Three angles are describes as follows: Yaw, Pitch, and Roll. In this, yaw angle (α) is estimated using below expression:

$$\alpha = \arccos(Z_3) \tag{17}$$

Where Z_2 and Z_3 represent the Z vectors of the given image.

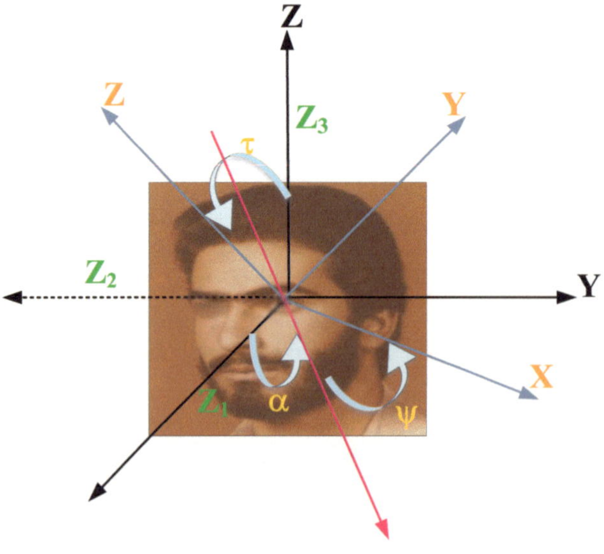

Figure 3.
Euler angles representation.

Roll angle (φ) can be estimated using below expression:

$$\varphi = \arccos\left(-\frac{Z_2}{\sqrt{1 - {Z_3}^2}}\right) \tag{18}$$

Pitch angle (τ) can be estimated using below expression:

$$\tau = \arccos\left(\frac{Y_3}{\sqrt{1 - {Z_3}^2}}\right) \tag{19}$$

These three angles are given as input to the affine transformation algorithm in order to rotate into the correct view. There exist four basic affine transformations that are illustrated as follows:

- Translate—It moves a set of point in fixed distance in x and y.
- Scale—It scales the set of points in up or down directions.
- Rotate—It rotates the set of points about the origin.
- Shear—It offsets a set of points in distance proportional to their x and y coordinates.

In mathematical form, an affine transformation of $\mathfrak{N}^n$ is a map of F: $\mathfrak{N}^n \rightarrow \mathfrak{N}^n$

$$F(\mathfrak{s}) = L_t(\mathfrak{s}) + \mathcal{Q}\forall\mathfrak{s} \in \mathfrak{N}^n \tag{20}$$

Where, L_t indicates the linear transformation of $\mathfrak{N}^n$, and $\mathcal{Q}$ defines the translation vector in $\mathfrak{N}^n$. A rotation performed in the affine transformation is illustrated as follows:

$$\text{Rotation about } x \text{ axis: } \begin{bmatrix} 1\ 0\ 0\ \ 0 \\ 0 \quad \cos\theta_x \ \ -\sin\theta_x \quad 0 \\ 0 \quad \sin\theta_x \ \ \cos\theta_x \quad 0 \\ 0\ \ 0\ 0\ \ 1 \end{bmatrix} \tag{21}$$

$$\text{Rotation about } y \text{ axis: } \begin{bmatrix} \cos\theta_x \quad 0\ \sin\theta_y \quad 0 \\ 0\ \ 0\ 0\ \ 0 \\ -\sin\theta_y \quad 0\ \cos\theta_y \quad 0 \\ 0\ \ 0\ 0\ \ 1 \end{bmatrix} \tag{22}$$

$$\text{Rotation about } z \text{ axis: } \begin{bmatrix} \cos\theta_z & -\sin\theta_z\ 0\ 0 \\ \sin\theta_z & \cos\theta_z \quad 0\ 0 \\ 0 \quad 0 & 1 \quad 0 \\ 0 \quad 0 & 0 \quad 1 \end{bmatrix} \tag{23}$$

Features	Feature description	Types of features
Texture	Texture feature represents the surface characteristics of the image	Contrast, Dissimilarity, Entropy, Homogeneity, Correlation, and Angular Second Moment
Shape	Shape features represents the physiological identity of given image	Boundary of the periocular, nose, and mouth regions, Convexity, and Solidity
Demographic	Demographic features represent the individual uniqueness of the given image.	Race, Age, and gender

Table 1.
Features description.

Where θ_x, θ_y, and θ_z represent the rotations about three axes known as Euler angles. This way of rotation in affine transformation results in frontal view of the given image. After completing pose normalization, we crop the image in order to change the size of all images into same one.

4.1.2.4 Feature extraction and fusion

Feature extraction and fusion are a major part of this work in order to produce optimum results in AIFR. Our MF-AIFR extracts multiple features from three set of regions. We extract images from three regions that are periocular, nose, and mouth. Since, these three regions are significant to recognize the image across aging. From these regions, we extract three type of features that are texture, shape, and demographic, which are briefed in **Table 1**. Here, texture feature is extracted using the CNN descriptor, and SIHKS descriptor is used to extract the shape and demographic-related features.

4.1.2.4.1 Texture feature extraction

Our MF-AIFR utilizes CNN descriptor for texture feature extraction since it provides robust performance in learning features layer by layer. CNN applies multiple filters on the raw input image in order to extract high-level features. Here, we extract six texture features in given image such as contrast, dissimilarity, entropy, homogeneity, correlation, and angular second moment. These features are described as follows: In CNN, three different types of layers are present that are Convolutional layer, Polling layer, and Fully connected layer.

4.1.2.4.2 Convolutional layer

It gathers image from the input layer, which is made up of a set of learnable filters. In our work, convolutional layer comprises six filters in order to generate feature map. Six filters in the convolutional layer generate six feature maps. The feature map is the consequence of the every filter that convolved through whole image. Convolution operation can be described as follows:

$$x_j^l = a_f\left(\sum_{i \in M_l} x_j^{l-1} * f_{ij}{}^l + \right) \tag{24}$$

Where a_f describes the activation function, j represents the specific convolution feature map, l represents the layer in the CNN, f_{ij} indicates the filter, b_j represents the feature map bias, and M_l is a selection of feature map.

4.1.2.4.3 Pooling layer

It is used to perform downsampling operation in order to reduce the spatial size of the convolutional layers. Polling operation is implemented on the pixel values captured by the pooling mask. The pooling operation is described as follows:

$$p^l_j = a_f\left(C^l_j pool\left(p^{l-1}_j\right) + b^l_j\right) \tag{25}$$

Where p^l_j represents the result of the pooling region applied on the jth region in the input image. p^{l-1}_j describes the jth region of interest captured by the pooling mask in previous layer. C^l_j indicates the trainable coefficient.

4.1.2.4.4 Fully connected layer

Fully connected layer is used to extract the features that are obtained in the preceding layers. The results obtained in the last convolutional and pooling layer are given as input to the fully connected layer in order extract features.

4.1.2.4.5 Shape and demographic feature extraction

Shape and demographic features are extracted using SIHKS algorithm. Shape features are boundary of the eye, nose and mouth, Convexity, and Solidity. Demographic features comprise age, race, and gender information. Here, race feature represents the skin tone of the face image. These features plays key role in recognizing face across aging.

Proposed SIHKS descriptor performs better than HKS algorithm since conventional method has drawback such as sensitivity to scale especially to the global scale. Hence, we proposed SIHKS algorithm, which performs better in scale invariance, and it is able perform at any point even at scale selection is impossible. In addition to it, it also performs well extracting shape and demographic-oriented features compared with other shape feature descriptor. SIHKS extracts features using three steps that are listed as follows:

- Logarithmical sampling in time t. It can be expressed using below equation.

$$h_t' = h'(x, \propto^t) \tag{26}$$

Where h_t' represents logarithm sampling of heat kernel signature.

- Taking logarithm of heat signature with time variations. It can be described as the below equation,

$$h_t' = h_{t+s} \text{With } h_t = \log h_{t+1} - h_t \tag{27}$$

Where h_{t+s} represent the shift in the heat kernel signature.

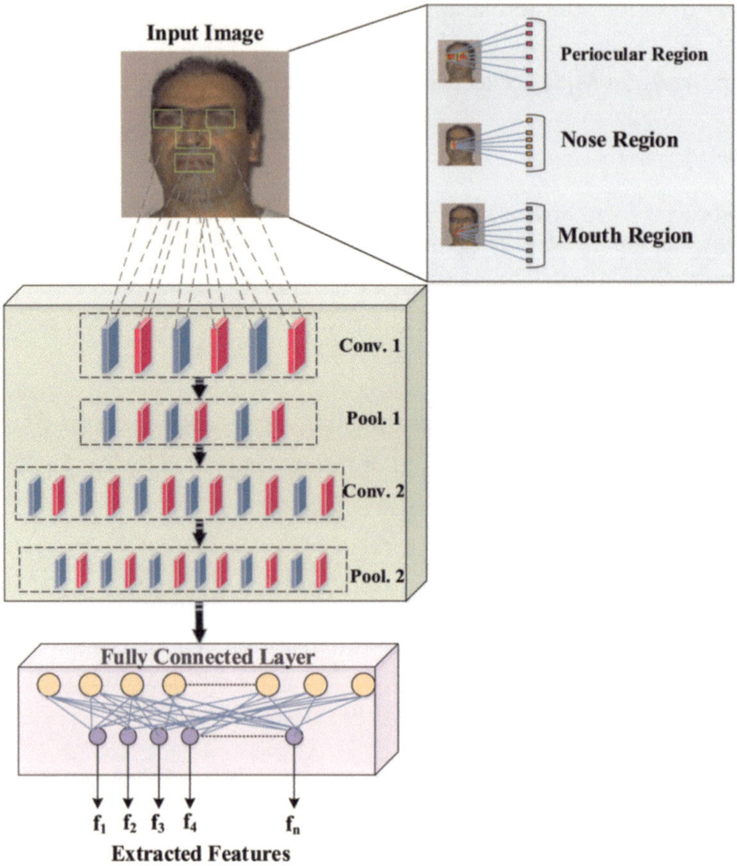

Figure 4.
Feature extraction in CNN.

- Taking discrete time Fourier transform of heat signature. It can be expressed as below equation,

$$F[h_t']w = H'[w] = H(w)e^{-2\pi ws} = F[h_{t+s}](w) \tag{28}$$

With the above steps, our SIKHS estimates scale-invariant quantity $|H(w)|$ at each point x without performing scale selection. Using this quantity, our SIKHS algorithm estimates the shape and demographic-oriented features effectually.

Figure 4 illustrates the texture feature extraction in CNN with their significant layers such as convolutional layer, pool layer, and fully connected layer.

4.1.2.4.6 Feature fusion

Feature fusion is estimated to reduce extracted feature dimension of extracted features such as shape, texture, and demographic features. This dimensionality reduction will result in better performance in face recognition, which the process of recognition and retrieval is easier. For this purpose, our MF-AIFR algorithm utilizes CCA algorithm, which performs effectively in feature fusion.

Feature fusion is defined as the combination of multiple feature vectors into single feature vector. Proposed CCA is a statistical tool for recognizing linear relationship among sets of features vectors in order to determine the inter subject covariances. Canonical covariates of the given feature vectors are obtained using below expression,

$$A_1{}^T = uX_1^T : A_2{}^T = vX_2^T : A_3{}^T = dX_3^T \tag{29}$$

Where A_1, A_2, A_3 represent the canonical covariates of the feature vectors $X_1, X_2, X_3,$ which indicates texture, shape, and demographic features. And, u, v, d describe the eigen vectors of the features.

4.1.3 Recognition and retrieval

Recognition and retrieval are final process in our MF-AIFR, which is performed by utilizing SVM algorithm. Here, we select SVM algorithm to correctly recognize the face cross aging and also retrieve the recognized image for given input image. **Figure 5** illustrates the input and output space models of the SVM algorithm.

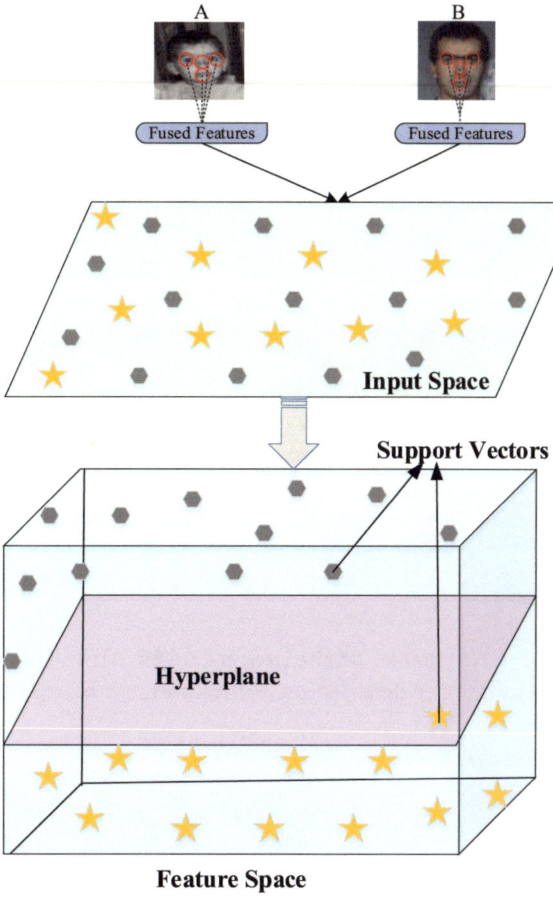

Figure 5.
SVM input and feature space representation.

Proposed SVM algorithm performs well in even unstructured and semistructured data. It addition to it, SVM also scales relatively well to high dimensionality of database. SVM gets input as fused features from previous process obtained using CCA algorithm. SVM is the binary classification method that discovers the optimal linear decision surface based on the concept of structural risk minimization. The decision surface represents the weighted combination of the elements present in the training set. These elements are illustrated as the support vectors and characterize the boundary between two different classes. The output of the SVM algorithm is a set of support vectors S_i, coefficient weights w_e, class labels y_i of the support vectors, and constant term z

The linear surface is represented as follows:

$$k.z + b = 0 \tag{30}$$

Where k represents the weight factor and b represents the bias term and z represents the training or testing data. These two parameters are used separate the hyperplane position and orientation. The weight factor k is calculated using below expression,

$$k = \sum_{i=1}^{N_s} w_{ei} y_i S_i \tag{31}$$

Kernel function plays vital role in SVM, which classifies features effectually. In MF-AIFR, we use Radial Basis Function (RBF) kernel. RBF performs well compared with other kernel functions. It doesn't require any prior knowledge about data. It can be expressed as follows:

$$r(v - v_i) = e^{-\delta \|v - v_i\|^2} \tag{32}$$

Here, δ represents the regularization parameter, and $v - v_i$ represents the different between feature vectors. By utilizing RBF kernel function, our MF-AIFR method recognizes and retrieves the images that are same as given test image. For example, if we give an input as image of person "A" at the age of 33, then it retrieves of the person "A" image from the age 2 to 60, since our FG-NET database contains subjects with the age from 0 to 69.

5. Experimental study

To characterize the performance of the proposed MF-AIFR, this section is divided into four aspects such as dataset description, simulation setup, application scenario, results, and discussion.

5.1 Dataset description

This section deliberates dataset information used in this chapter. Here, we utilize FG-NET database to perform face recognition and retrieval. Face and gesture recognition NETwork (FG-NET) aging database was released in the year of 2004 in an attempt to support research activities regarding the changes in the facial appearance

Parameters	Values	# Images
# subjects	82	1002
#Males	34	Max (1–12) per subject
#Females	48	Max (1–12) per subject

Table 2.
Dataset description.

Factors	Ages									
	0–5	6–10	11–15	16–20	21–25	26–30	31–35	36–40	41–45	46–69
#Subjects	75	70	71	68	46	38	30	24	19	10
#Images	233	178	164	155	81	62	38	31	26	34

Table 3.
Different age bands of FG-NET dataset.

caused by aging. FG-NET database comprises 1002 images from 82 different subjects. Each subject comprises 6–18 images with the age ranging between the newborns to the 69-year-old subjects. Our FG-NET database contains considerable variations such as poses and illuminations.

Table 2 illustrates the details of the FG-NET dataset briefly. Dataset contains 34 male subjects and 48 female subjects' images. Each subject has 1–12 images across their age progression.

Different age bands present in the FG-NET dataset are represented in **Table 3**. FG-NET dataset comprises subjects from the age of 0 to 69 years old.

5.2 Experimental setup

Our proposed MF-AIFR is implemented in MATLAB R2017b tool with C programming language. Our MATLABR2017b is executed in windows operating system. MATLAB is a multi-paradigm statistical computing environment developed by MathWorks. MATLAB permits matrix manipulations, implementation of algorithms plotting of functions and data, creation of user interfaces, and interfacing with programs written in other languages, which include C, C++, C#, JAVA, and Python.

5.3 Performance metrics

To evaluate performance of the MF-AIFR, we consider following metrics that are described as follows:

- *Accuracy:* It is defined as the ratio of correct classification with respect to that of total images. The accuracy is measured based on the succeeding expression:

$$Accuracy = \frac{\#of\ correct\ classification}{Total\ images} \tag{33}$$

- *Recall:* It is defined as the proportions of the cases that are correctly classified by a class. Recall is also called as True Positive (TP) cases. It can be illustrated as follows:

$$Recall = \frac{T_P}{T_P + F_N} \tag{34}$$

Where T_P is defined as the positive cases that are correctly labeled as positive. F_N is defined as the noise samples that are incorrectly labeled as negative.

- *Precision:* It is designated as the ratio of the number correctly classified samples with all classified samples. Precision is also called as positive predictive value. It can be designated as follows:

$$Precision = \frac{T_P}{T_P + F_p} \tag{35}$$

Where F_p represents the number of noise lesions correctly detected as samples.

- *F-Score:* It is the combination of precision and recall. It can be calculated as follows:

$$F - Score = \frac{2 * (Recall * Precision)}{Recall + Precision} \tag{36}$$

- *Recognition Rate:* It is evaluated based on the features of the face image. It describes the face recognition ability in AIFR.
- *Rank 1 Score*: It represents the Cumulative Matching Result (CMR) for given image. It can be used to detect the correctly matched score for given FG-NET database images.
- *Computation Time:* It is designated as the total time required to retrieve the images for given input. This metric illustrates the efficacy of the proposed work in terms of time.

5.4 Comparative analysis

This compares the simulation results of the MF-AIFR with existing methods such as HLD, DF, and CAN. Here, we compare results using six performance metrics that are Accuracy, Recall, Precision, Recognition Rate, Rank-1 Score, and F-Score. **Table 4** illustrates the comparisons of previous methods with their strength, weakness, and research statements.

5.4.1 Impact on accuracy

Accuracy metric is one of the significant metrics to evaluate the performance of the proposed work. This metric defines the how accurate our MF-AIFR in terms of correct classification of images. The performance of this metric is evaluated by alternating the number of images.

Reference	Key concentration	Strength	Weakness	Research statements					
				Accuracy	Recall	Precision	Recognition rate	F-Score	Rank 1-score
Kishore et al. [41]	HLD-AIFR & Retrieval	Adopts large datasets	It removes fine details of the image and resultant image is blurred. Feature extraction loses more information due to lack of pixel consideration.	Low	Medium	Low	Low	Medium	Low
Muhammad et al. [42]	DF-AIFR & Retrieval	Better demographic Estimation	Takes more time in feature extraction	Very Low	Low	Low	Very Low	Low	Very Low
Chenfei et al. [13]	CAN-AIFR	Complexity is less	Data relationships are not considered that affects the recognition results.	Low	Medium	Very Low	Low	Low	Low
Huiling et al. [15]	IIM-AIFR	Flexible to large dataset	More noise in extracted features due to absence of noise removal	Medium	Low	Very Low	Medium	Very Low	Low
Fahad et al. [9]	CTS-AIFR	Recognition time is less	Naïve Bayes based recognition results are always biased one, since it doesn't rely on class conditional dependency.	Very Low	Medium	Very Low	Very Low	Very Low	Low

Table 4.
Comparisons on previous methods in AIFR.

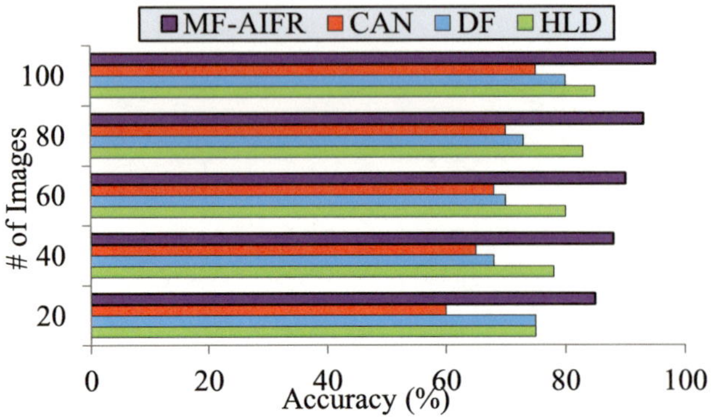

Figure 6.
Comparisons on accuracy.

Figure 6 demonstrates that comparisons on accuracy of the MF-AIFR with respect to the existing methods such as CAN, DF, and HLD. These comparisons show that our MF-AIFR achieves better performance compared with the existing methods. Since, our method utilizes better feature descriptors such as CNN and SIHKS. Both algorithms extract features effectually from three regions that are periocular, nose, and mouth. This selected region plays a key role in recognizing face across aging. And CNN and SIHKS provide robust performance even in high-dimensional dataset. As a result, our method achieves high accuracy as 95%. By contrast, CAN and DF method attain less accuracy compared with our method due to its poor feature extraction procedures since it doesn't concentrate on the vital regions such as periocular, nose, and mouth. Meanwhile, HLD obtains high accuracy compared with both CAN and DF method due to its feature extraction from periocular region, which plays significant role in face recognition across aging. Though, it achieves less accuracy compared with our method due to its poor descriptor algorithm since it loses large amount of information during feature extraction.

Table 5 illustrates the average simulation results comparison of accuracy with the existing and proposed methods.

From the above comparison, it is noticed that our method achieves better accuracy percentage as 90.2% compared with the existing methods.

5.4.2 Impact on recall

Recall is used to evaluate the performance of the MF-AIFR in terms of the correct recognition of face image. Recall performance is evaluated by changing the number of images.

Methods	**Accuracy (%)**
HLD	80.2
DF	73.2
CAN	67.6
MF-AIFR	90.2

Table 5.
Accuracy comparisons [average].

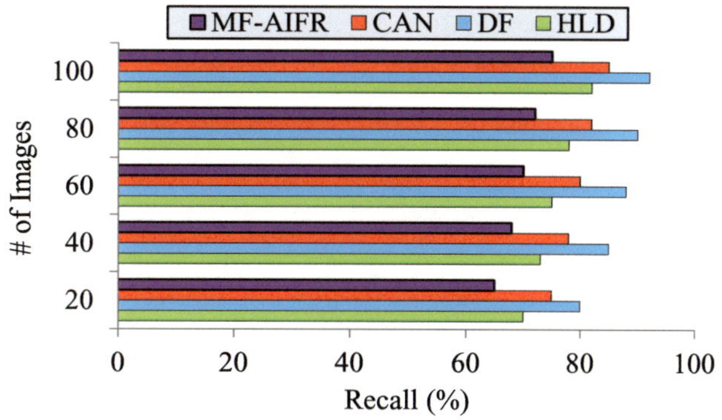

Figure 7.
Comparisons on recall.

Methods	Recall (%)
HLD	75.6
DF	87
CAN	80
MF-AIFR	70

Table 6.
Recall comparisons [average].

Figure 7 shows that our MF-AIFR achieves less recall percentage compared with other methods.

Since, our MF-AIFR correctly recognizes the face as per given test image, thus reduces false detection of face images. Reason for this is that our method executes pose normalization before entering into the feature extraction process. Pose normalization enhances the feature extraction efficiency. Thus it leads to correct identification and retrieval of the test image. As a result, our MF-AIFR achieves less recall percentages compared with existing methods. Whereas existing methods such as DF and CAN achieves high recall percentages due to lack of pose normalization and complex feature extraction procedures. In the meantime, HLD method reduces recall percentage compared with DF and CAN methods since it doesn't follow complex feature extraction procedures. Still, recall of HLD is high compared with MF-AIFR due to lack of pose normalization and information degradation in noise removal process. **Table 6** designates the average simulation results comparison of recall with the existing and proposed methods.

From the above comparison results, it is seen that our MF-AIFR method achieves less recall percentage as 70% compared with the existing methods.

5.4.3 Impact on precision

Precision is used to measure performance of our work in terms of relevance instances retrieved compared with the total images. Precision performance is measured via altering the number of image.

Figure 8 depicts that MF-AIFR achieves high precision percentages compared with existing methods. MF-AIFR performs preprocessing process before entering into the feature extraction and recognition process. Preprocessing performs illumination

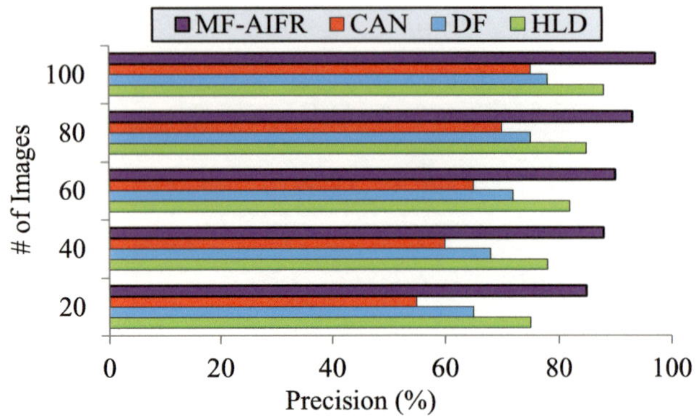

Figure 8.
Comparisons on precision.

Methods	Precision (%)
HLD	81.6
DF	71.6
CAN	65
MF-AIFR	90.6

Table 7.
Precision comparisons [average].

normalization and noise filtering since our FG-NET dataset contains illumination and noises in images. These two processes enhance the quality of the image that tends to easy the feature extraction and recognition process. CAN and DF methods achieves less precision due to lack of preprocessing such as noise removal and illumination normalization. Likewise, HLD also obtains less precision owing to fine detail removal in Gaussian-based noise filtering. Since Gaussian filter doesn't concentrate on fine details of the image, which results in blur image.

Table 7 designates the average simulation results comparison of precision with the existing and proposed methods. From the above comparison, we conclude that MF-AIFR achieves better precision percentage as 90.6% compared with existing methods.

5.4.4 Impact on F-score

F-Score metric considers both false positive and false negative values in account to estimate performance of this work. The performance of this metric is simulated by varying the number of images.

Figure 9 illustrates that comparison on F-Score result of MF-AIFR with existing methods such as DF, CAN, and HLD. From this figure, it is noticed that our method achieves high F-Score compared with existing methods. Our MF-AIFR uses two descriptors such as CNN and SIHKS to extract texture, shape, and demographic features. Here, SIHKS descriptor performs very well in scale invariance and also provides better extraction results even when scale selection is impossible. It extracts shape and demographic features effectually, which plays substantial role in face recognition across aging. At the same time, CAN and DF methods attain less F-Score owing to the absence

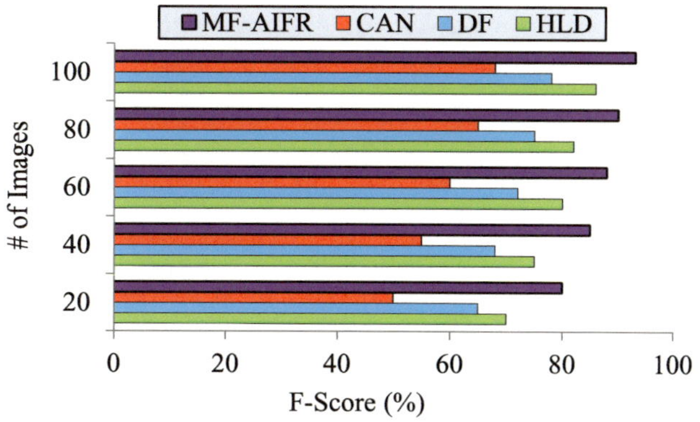

Figure 9.
Comparisons on F-score.

Methods	F-Score (%)
HLD	78.6
DF	71.6
CAN	59.6
MF-AIFR	87.2

Table 8.
F-score comparisons [average].

of significant feature extraction such as texture and shape features. Meanwhile, HLD also attains less F-Score since it doesn't concentrate on shape features extraction and thus reduces the face recognition and retrieval efficiency.

Table 8 describes the average simulation results comparison of F-Score with the existing and proposed methods. From the above comparison, we observed that MF-AIFR method achieves high F-Score percentage as 87.2% compared with existing methods.

5.4.5 Impact on recognition rate

Recognition rate is used to measure the ability of MF-AIFR in terms of the face recognition. It can be measured through changing the number of features.

Figure 10 designates the comparisons on recognition rate of MF-AIFR with respect to the existing methods CAN, DF, and HLD methods. From this figure, it is observed that our MF-AIFR attains high recognition rate compared with existing method. We propose SVM algorithm for recognition and retrieval. It performs well in recognition even in high dimensionality of dataset. In addition to it, we also perform feature fusion before entering into the recognition and retrieval process.

Feature fusion reduces the dimension of feature vectors and thus tends to enhance the performance of SVM algorithm. Therefore, our method achieves better recognition rate compared with existing method. Meanwhile, DF method has less recognition rate compared with other methods due to lack of effective recognition and retrieval processes since it simply ranks the images. Likewise, CAN also attains less recognition rate compared with our method since it isn't able to establish data relationship between different features. Meantime, HLD method attains less recognition rate due

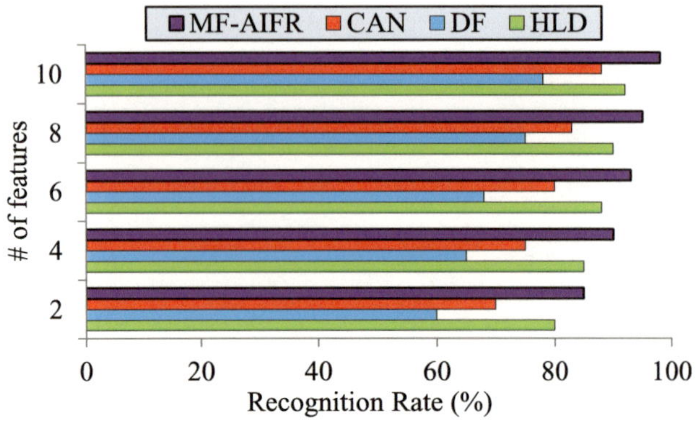

Figure 10.
Comparisons on recognition rate.

Methods	Recognition rate (%)
HLD	87
DF	69.2
CAN	79.2
MF-AIFR	92.2

Table 9.
Recognition rate comparisons [average].

to usage of KNN for recognition. KNN takes more time, and discovering similarity measure is tedious.

Table 9 defines the average simulation results comparison of recognition rate with the existing and proposed methods. Above comparison illustrates that recognition rate of MF-AIFR is higher than that of other existing methods.

5.4.6 Impact on rank-1-score

Rank-1 Score considers the performance of cumulative match for given images in proposed work. It represents the efficacy of our work in terms of recognition and retrieval.

Figure 11 exhibits comparisons on rank-1 score results with respect to the existing methods. From this figure, it is seen that our MF-AIFR attains high rank-1 score compared with the existing methods. Our proposed DGC-CLAHE algorithm based illumination normalization performs well compared with existing CLAHE; it enhances the fine details of the image. ASBF-based noise filtering also provides better performance in noise removal, which sharpens the image. This way of preprocessing results in high matching results in face recognition. At the same time, existing methods such as DF and CAN attain less rank 1 score since it doesn't use effective algorithm for preprocessing and thus reduce the quality of given image drastically. Likewise, HLD also attains less rank 1 score compared with our method. Since, it doesn't perform illumination normalization and noise filtering also not effective. From this analysis, we conclude that our MF-AIFR attains better results in rank 1-score compared with other methods.

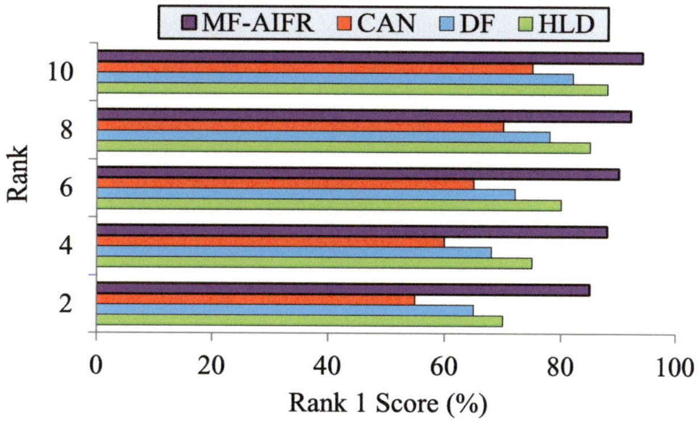

Figure 11.
Comparisons on rank 1-score.

Methods	Rank 1 score
HLD	79.6
DF	73
CAN	65
MF-AIFR	89.8

Table 10.
Rank 1 score comparisons [average].

Table 10 signifies average simulation results comparison of rank 1-score with the existing and proposed methods. From the above comparison, we prove that our MF-AIFR method achieves higher rank 1 score percentage as 89.8% compared with existing methods.

5.4.7 Impact on computation time

Performance of the computation time is evaluated by varying the number of images. This metric must be low in order to attain better performance in image retrieval across aging.

Figure 12 depicts the comparisons on computation time results with respect to the existing methods. It is noticed that our MF-AIFR method achieves less computation time compared with the existing methods such as CAN, DF, and HLD. MF-AIFR performs IQE process before entering into the preprocessing step. The images that are not satisfying IQT only undergone preprocessing; otherwise it is directly given to the pose normalization step. Thus it reduces the time wastages in performing preprocessing for all input images. In addition to it, our work also reduces time in feature extraction and classification by using effective algorithms such as CNN, SIHKS, and SVM. These algorithms require less time to process the given inputs. As a result, MF-AIFR achieves less computation time. In the meantime, existing methods such as CAN and DF attain high computation time compared with other methods. Since it performs preprocessing for all images and also doesn't utilize effective algorithm to process the given input image and thus leads to increase in computation time. Likewise, HLD also attains high computation time compared with MF-AIFR since it performs preprocessing for all images regardless of their quality.

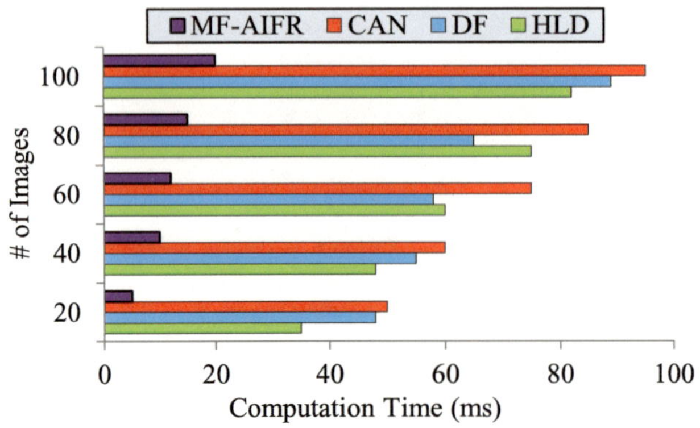

Figure 12.
Comparisons on computation time.

Methods	Computation time (ms)
HLD	60
DF	63
CAN	73
MF-AIFR	12.4

Table 11.
Computation time comparisons [average].

Table 11 deliberates the comparisons of computation time and thus shows that our method attains less computation time as 12.4ms compared with other methods including HLD, DF, and CAN.

5.5 Research highlights

This section signifies highlights of this research regarding face recognition across aging. In order to achieve better performance in AIFR, our work establishes five consequent processes. **Table 12** describes the benefits of proposed algorithms along

Algorithms	Main functionality	Benefits related to performance
DGC-CLAHE	Illumination Normalization	Enhances the recognition rate and accuracy
ASBF	Noise removal	Enhances recognition rate and feature extraction efficiency
EA-AT	Pose Normalization	Easier the feature extraction process and Increases the precision level
CNN	Texture Feature Extraction	Enhances the accuracy in face recognition across aging and perform well in large scale data set
SIHKS	Shape & Demographic Feature extraction	Increases the rank 1-score and adapts large scale data set.
SVM	Recognition and Interval	Simple processing, increases the accuracy and reduces the recall

Table 12.
Benefits of proposed algorithms.

with their functionalities. This table illustrates each algorithm with their benefits in performance metrics such as precision, recall, accuracy, recognition rate, and rank 1 score.

6. Conclusion and future work

Face recognition across aging becomes challenging due to changes in the human faces with age progressions. In order to address this bottleneck, this chapter proposes MF-AIFR method where four successive processes performed that are listed as follows: IQE is performed to reduce time spend in preprocessing and thus enhances performance of our system drastically. An image that doesn't satisfy the IQT is given as input to the preprocessing step. Here, illumination normalization and noise removal are performed, which enhances the accuracy in face recognition and retrieval. Illumination normalization adopts DGC-CLAHE, and noise removal adopts ASBF algorithm. In order to normalize the pose, we adopt EA-AT algorithm, which is performed to enhance the feature extraction efficacy. Two types of descriptors are utilized for features extractions that are CNN and SIHKS. Here, we extract multiple features such as texture, shape, and demographic features. We extract features from three types of regions that are periocular, nose, and mouth. CNN extracts texture features, and SIHKS extracts shape and demographic features. This way extracting features increases our recognition rate. In recognition and retrieval, we execute SVM algorithm, which follows the simple procedure and provides better results. At last, we evaluate the performance of MF-AIFR system using seven metrics that are Accuracy, Recall, Precision, Rank-1 Score, F-Score, Recognition rate, and Computation time. Thus it shows that our work performs better than existing methods such as HLD, DF, and CAN.

Author details

Kishore Kumar Kamarajugadda* and Movva Pavani
Faculty of Science and Technology, Department of Electronics and Communication Engineering, ICFAI Foundation for Higher Education, Hyderabad, India

*Address all correspondence to: kkishore@ifheindia.org

References

[1] Tong SG, Huang YY, Tong ZM. A robust face recognition method combining LBP with multi-mirror symmetry for images with various face interferences. International Journal of Automation and Computing. 2019;**16**: 1-12. DOI: 10.1007/s11633-018-1153-8

[2] Moon HM, Seo CH, Pan SB. A face recognition system based on convolution neural network using multiple distance face. Soft Computing. 2017;**21**(17): 4995-5002

[3] Roh S-B, Oh S-K, Yoon J-H, Seo K. Design of face recognition system based on fuzzy transform and radial basis function neural networks. Soft Computing. 2019;**23**(13):4969-4985

[4] Singh R, Om H. New born face recognition using deep convolutional neural network. Multimedia Tools and Applications. 2018;**76**(18):19005-19015

[5] Agarwal V, Bhanot S. Radial basis function neural network based face recognition using firefly algorithm. Neural Computing and Applications. 2018;**30**(8):2643-2660

[6] Wang Y, Gong D, Zheng Z, Ji X, Wang H, Li Z, et al. Orthogonal deep features decomposition for age-invariant face recognition. Computer Vision and Pattern Recognition. 2018:1-13

[7] Wang H, Gong D, Li Z, Liu W, Tencent AI Lab. Decorrelated adversarial learning for age-invariant face recognition. Computer Vision and Pattern Recognition. 2019:1-10

[8] Li Z, Gong D, Li X, Tao D. Aging face recognition: a hierarchical learning model based on local patterns selection. IEEE Transactions on Image Processing. 2016;**25**(5):2146-2154

[9] Kishore KK, Trinatha RP. Biometric identification using the periocular region, information and communication technology for intelligent systems (ICTIS 2017), volume 2. Smart Innovation, Systems and Technologies. Cham: Springer. 2018;**84**:619-628. DOI: 10.1007/978-3-319-63645-0_69

[10] Sawant MM, Bhurchandi K. Age invariant face recognition: A survey on facial aging databases, techniques and effect of aging. Artificial Intelligence Review. 2018:1-28

[11] Feng S, Lang C, Feng J, Wang T, Luo J. Human facial age estimation by cost-sensitive label ranking and trace norm regularization. IEEE Trans Multimedia. 2017;**19**(1):136-148

[12] Dhamija A, Dubey RB. Analysis on age invariance face recognition study and effects of intrinsic and extrinsic factors on skin ageing. International Journal of Computer Applications. 2019; **182**(43):1-9

[13] Kishore KK, Trinatha RP. Face verification across ages using discriminative methods and see 5.0 classifier. In: Proceeding of First International Conference on Information and Communication Technology for Intelligent Systems: Volume 2. Smart Innovation, Systems and Technologies, vol 51. Cham: Springer. 2016. pp 439–448. DOI: 10.1007/978-3-319-30927-9_43

[14] Li Z, Park U, Jain AK. A discriminative model for age invariant face recognition. IEEE Transactions on Information Forensics and Security. 2011;**6**:1028-1037

[15] Kishore KK, Trinatha RP. Periocular region based biometric identification

using the local descriptors. Intelligent Computing and Information and Communication. Advances in Intelligent Systems and Computing. Singapore: Springer. 2018;**673**:341-351. DOI: 10.1007/978-981-10-7245-1_34

[16] Kamarajugadda KK, Polipalli TR. Age-invariant face recognition using multiple descriptors along with modified dimensionality reduction approach. Multimedia Tools and Applications. 2019;**78**(19):27639-27661. DOI: 10.1007/s11042-019-7741-y

[17] El Khiyari H, Wechsler H. Age invariant face recognition using convolutional neural networks and set distances. Journal of Information Security. 2017;**8**:174-185

[18] Divyanshu S, Pandey JP, Chauhan B. A deep learning approach for age invariant face recognition. International Journal of Pure and Applied Mathematics. 2017;**117**(21):371-389

[19] Bianco S. Large age-gap face verification by feature injection in deep networks. Pattern Recognition Letters. 2017;**90**:36-42

[20] Riaz S, Ali Z, Park U, Choi J, Masi I, Natarajan P. Age-invariant face recognition using gender specific 3D aging modeling. Multimedia Tools and Applications. 2019:1-21

[21] Kumar KK, Pavani M. Periocular region-based age-invariant face recognition using local binary pattern. Microelectronics, Electromagnetics and Telecommunications. Lecture Notes in Electrical Engineering. Singapore: Springer. 2019;**521**:713-720. DOI: 10.1007/978-981-13-1906-8_72

[22] Kishore Kumar K, Pavani M. LBP based biometrie identification using the periocular region. In: 8th IEEE Annual Information Technology, Electronics and Mobile Communication Conference (IEMCON). Vancouver, BC: IEEE; 2017. pp. 204-209. DOI: 10.1109/IEMCON.2017.8117193

[23] Kamarajugadda KK, Polipalli TR. Extract features from periocular region to identify the age using machine learning algorithms. Journal of Medical Systems, Springer. 2019;**43**:196. DOI: 10.1007/s10916-019-1335-0

[24] Nanni L, Lumini A, Brahnam S. Ensemble of texture descriptors for face recognition obtained by varying feature transforms and preprocessing approaches. Applied Soft Computing. 2017;**61**:8-16

[25] Nhan Duong C, Gia Quach K, Luu K, Le N, Savvides M. Temporal non-volume preserving approach to facial age-progression and age-invariant face recognition. IEEE International Conference on Computer Vision (ICCV). 2017:3755-3763

[26] Chen B-C, Chen C-S, Hsu WH. Cross-age reference coding for age-invariant face recognition and retrieval. European Conference on Computer Vision ECCV 2014: Computer Vision—ECCV 2014. 2014:768-783

[27] Li Y, Wang G, Nie L, Wang Q, Tan W. Distance metric optimization driven convolutional neural network for age invariant face recognition. Pattern Recognition. 2018;**75**:51-62

[28] Chandran PS, Byju NB, Deepak RU, Nishakumari KN, Devanand P, Sasi PM. Missing child identification system using deep learning and multiclass SVM. IEEE Recent Advances in Intelligent Computational Systems (RAICS). 2018

[29] Verma G, Jindal A, Gupta S, Kaur L. A technique for face verification across

age progression with large age gap. In: 2017 4th International Conference on Signal Processing, Computing and Control (ISPCC). Solan, India: IEEE. 2017

[30] Bijarnia S, Singh P. Pyramid Binary Pattern for Age Invariant Face Verification. In: 2017 13th International Conference on Signal-Image Technology & Internet-Based Systems (SITIS). Jaipur, India: IEEE. 2017

[31] Nimbarte M, Bhoyar KK. Face recognition across aging using GLBP features. Information and Communication Technology for Intelligent Systems (ICTIS 2017). Smart Innovation, Systems and Technologies. Cham: Springer. 2017;**84**:275-283. DOI: 10.1007/978-3-319-63645-0_30

[32] Xu Z, Jiang Y, Wang Y, Zhou Y, Li W, Liao Q. Local polynomial contrast binary patterns for face recognition. Neuro Computing. 2019;**355**:1-12

[33] Mohanraj V, Sibi Chakkaravarthy S, Vaidehi V. Ensemble of convolutional neural networks for face recognition. Advances in Intelligent Systems and Computing Recent Developments in Machine Learning and Data Analytics. 2019:467-477. DOI: 10.1007/978-981-13-1280-9_43

[34] Kute RS, Vyas V, Anuse A. Component-based face recognition under transfer learning for forensic applications. Information Sciences. 2019; **476**:176-191

[35] Venkata Kranthi B, Surekha B. Real-time facial recognition using deep learning and local binary patterns. Proceedings of International Ethical Hacking Conference. 2018;**2018**:331-347

[36] Malek ME, Azimifar Z, Boostani R. Age-based human face image retrieval using zernike moments. Artificial Intelligence and Signal Processing Conference (AISP). 2017:347-351

[37] Wang D, Cui Z, Ding H, Yan S, Xie Z. Face aging synthesis application based on feature fusion. In: 2018 International Conference on Audio, Language and Image Processing (ICALIP). Shanghai, China: IEEE. 2018

[38] Kamarajugadda KK, Polipalli TR. Stride towards aging problem in face recognition by applying hybrid local feature descriptors. Evolving Systems. 2018;**10**:689–705. DOI: 10.1007/s12530-018-9256-6

[39] Shafique MST, Manzoor S, Iqbal F, Talal H, Qureshi US, Riaz I. Demographic-assisted age-invariant face recognition and retrieval. Symmetry. 2018;**10**(5):1-17

[40] Xu C, Liu Q, Ye M. Age invariant face recognition and retrieval by coupled auto-encoder networks. Neurocomputing. 2017;**222**:62-71

[41] Zhou H, Lam K-M. Age-invariant face recognition based on identity inference from appearance age. Pattern Recognition. 2018;**76**:191-202

[42] Alvi FB, Pears R. A composite spatio-temporal modeling approach for age invariant face recognition. Expert Systems With Applications. 2017;**72**: 383-394